T3-BUK-635

PRINTED CIRCUIT BOARDS

Revised by

Neal A. Willison
Qtek Consultant

Qtek Support Team:
Russell L. Heiserman, Technical Editor
John Jagoe, Language Editor
Mary Snavely, Manuscript Consultant

John Wiley & Sons
New York Chichester
Brisbane Toronto Singapore

Library of Congress Cataloging in Publication Data:

Willison, Neal A.
Printed circuit boards.

Includes index.
1. Printed circuits. I. Heiserman, Russell L.
II. Jagoe, John, 1952– III. QTEK (Company)
IV. Title.
TK7868.P7W54 1983 621.381'74 82-13541
ISBN 0-471-86177-4

Printed in the United States of America

10 9 8 7 6 5 4 3 2 1

Preface

In 1967, the Engineering Industry Training Board (EITB) in Great Britain developed a series of manuals and introduced new techniques designed to improve the quality of the training of craftsmen and technicians in the British industry. Since then, experience has guided revisions of and improvements in the format, the content, and the organization of the original manuals.

In 1980, John Wiley and Sons invited a team of experts representing U.S. industrial training and public vocational education to evaluate recent editions of the EITB manuals. The team's charge was to identify those materials most appropriate for U.S. training programs and make recommendations that would enhance the effectiveness of the selected materials. The team reacted to the material with enthusiasm and offered excellent suggestions for reworking the materials for the U.S. market.

Following this review, QTEK, a company that specializes in developing training materials, was selected to carry out the recommendations of the review team. QTEK was selected because it is composed of very practical, technical authors. All of the QTEK consultants who worked on this project are published authors and are members of the teaching faculty in the School of Technology at Oklahoma State University.

The result of this cooperative effort is a set of manuals unique in format, style, and training effectiveness. Each manual deals with a group of related tasks that a technician would be expected to know and be able to perform. The tasks are thoroughly covered in well-illustrated modules that emphasize getting the job done properly. At the end of each module is a quiz to help the student assess his or her progress and initiate a review of the material when necessary.

The manuals are suited for individual or small-group use. In an industrial setting, the user should be paired with an experienced technician acting as mentor in relating the manual's contents to specific needs of the company and in helping the trainee to locate and become familiar with proper use of specialized tools and instruments.

In vocational schools, the teacher will be able to make the material more meaningful through demonstrations and shop work. After the students become familiar with their manual or manuals, selected plant visits should be made, to reinforce the concept that these manuals do lead to knowledge and job skills that are practical, important, and needed.

Introduction

Your instructor has chosen class-tested training materials for you to use. These materials are organized as a series of modules to help you effectively develop skills and knowledge for doing specific jobs.

The way to get the most from your manual is to first become familiar with its general content. Thumb through the manual to get an overview of content and format. Next, read through the first module and think through the test items at the end of the module. Write in your answers. Each question has a reference to where the material is covered in the module—take just a moment to re-read the referenced material. Often, re-reading specific items will help you remember key ideas.

As you work through the manual try to relate the material to a company's equipment and jobs you might perform. It is helpful to work with an experienced technician and to ask questions. The manual can save you a lot of time and help you do a good job correctly.

The publishers and revisors of this material are interested in your comments, and recommendations. After all, it was developed for you. Please send your suggestions and recommended improvements to—The Occupational Publishing Group, John Wiley & Sons, 605 Third Ave, New York, New York 10158.

Contents

PRINTED CIRCUIT BOARDS

Module one
Printed circuit techniques

Introduction

Printed circuit boards are widely used in today's electronic circuits. The techniques used in manufacturing printed circuit boards will be examined in this module.

The manufacturing process begins with a blank printed circuit board or card. An artwork master of the circuit layout is then developed. The final stage requires an etching process to remove any unwanted conductive material.

Key technical words

- **Printed circuit board** A metal foil conducting pattern supported by an insulating substrate.
- **Silk screening** A method of transferring the conducting pattern to the copper board.
- **Etching** The chemical removal of unwanted copper.

Performance objectives

Upon completion of this module, you should be able to:

- Describe the silk screen method used to manufacture printed circuit boards.
- Describe the photographic method used to manufacture circuit boards.
- List the most widely used chemicals for etching printed circuit boards.

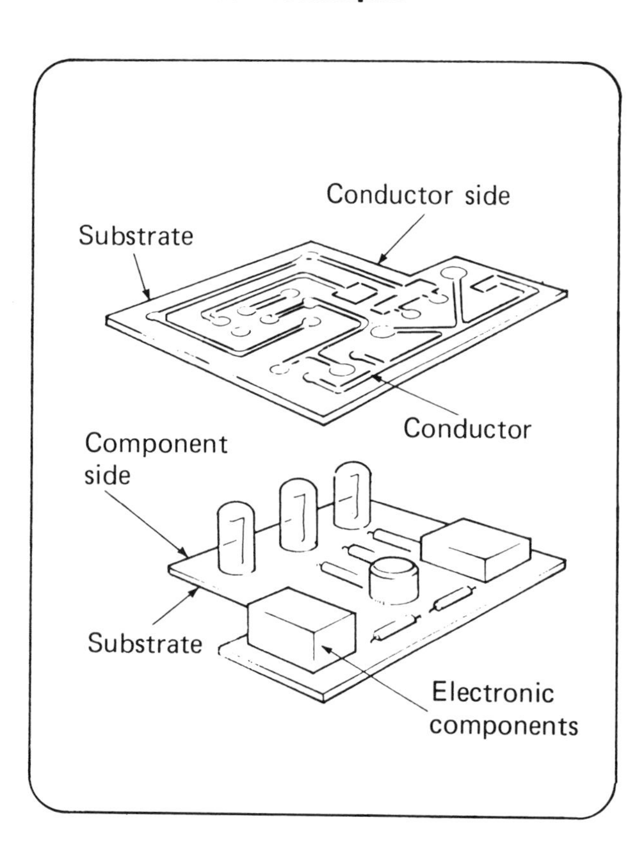

Printed circuit boards— Single-sided

A single-sided printed circuit board consists of a metal foil conducting pattern, usually copper, bonded to one surface of a substrate or insulating base material for support. The metal foil conducting pattern, also called lands or conductors, makes the interconnections between electronic components form an electronic circuit. The printed circuit board's sides are referred to as the component side, the side on which the electronic components are placed, and the conductor side.

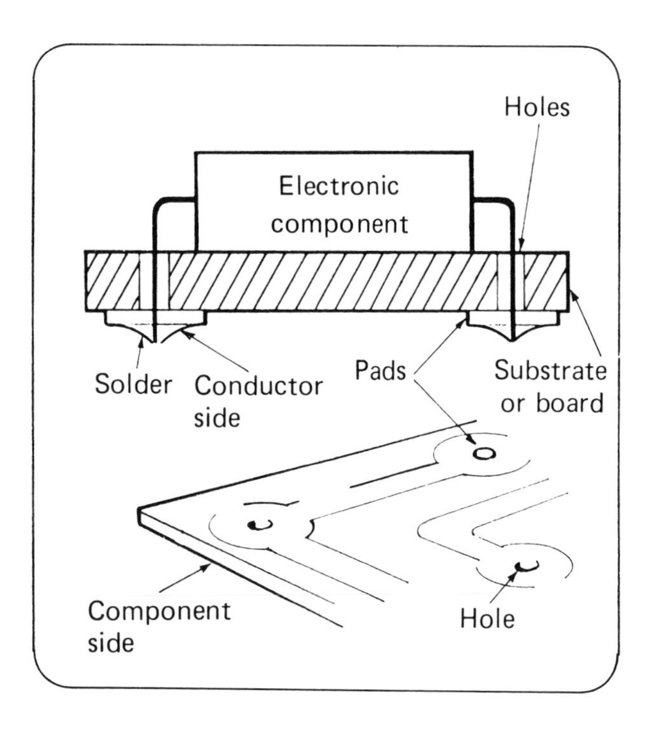

The components are mounted to the side of the board opposite to the conductor pattern. The board provides support for the components on one side and support for the metal foil on the other side.

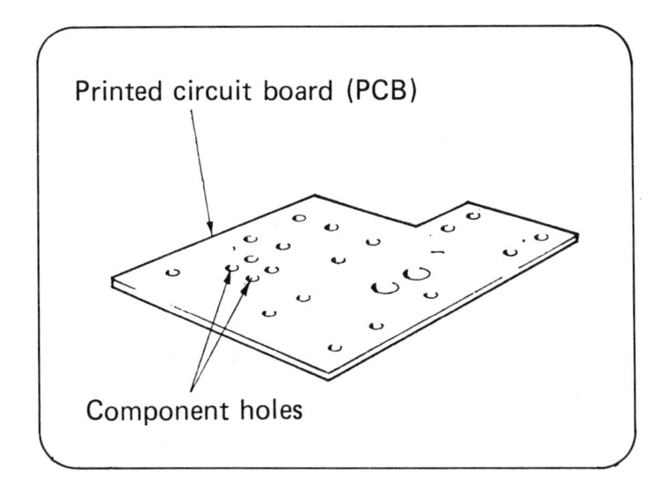

Component wires or leads pass through holes in the board and are soldered to the conductor patterns at points called pads.

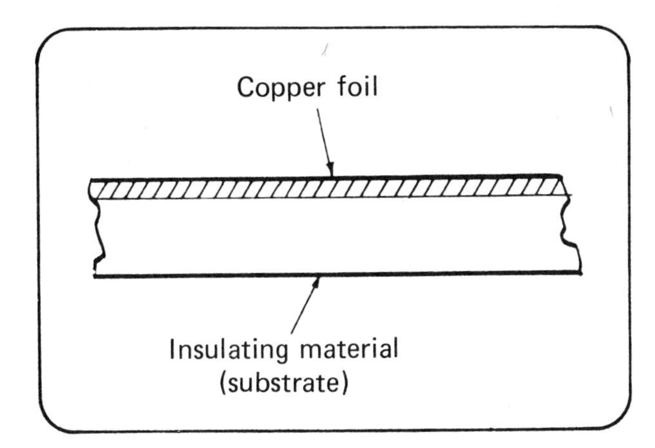

The printed circuit board

The laminated board, called a printed circuit board (PCB), is used to describe the total board: the substrate and the metal foil conductor. The board consists of a metal foil, usually copper, laminated to an insulating material or substrate. There are many different insulating materials used for the board. They are classified into two major groups.

1 Resin bonded paper–phenolic

2 Epoxy resins

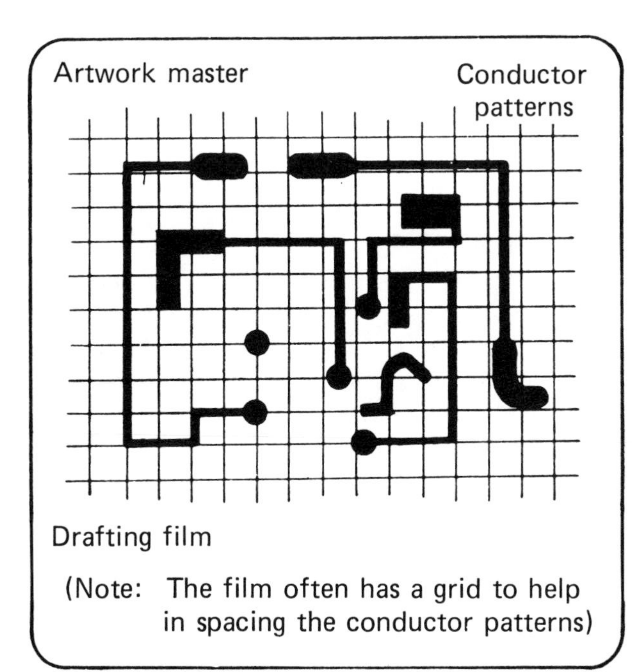

The artwork master

The artwork master provides the correct electrical connections for the circuit and the correct spacing between components. It is made by laying out the conductor pattern on a specialized drafting film by using printed circuit drafting tape and drafting patterns.

The small size of many printed circuit boards may require that the artwork be laid out larger than the actual size. The artwork is then photographically reduced to the correct size.

Methods of manufacture

To make a printed circuit board, the manufacturer removes the unwanted copper foil from the laminate, which leaves only the required pattern of copper foil. The areas of the copper foil that are to form the conductors are covered with a special chemical. This chemical, known as resist, protects the foil underneath from the etching process. The function of the etching process is to remove the excess copper.

There are two methods used to manufacture printed circuit boards.

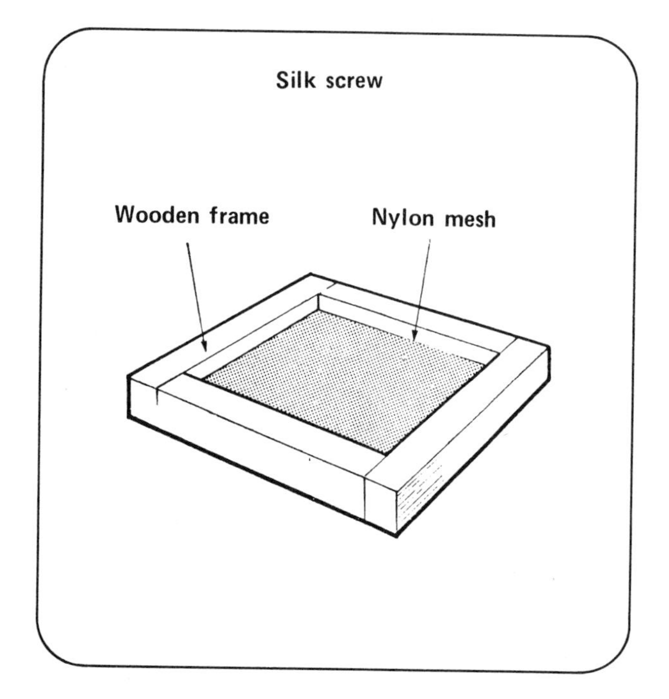

The silk screen method

The silk screen method is one technique of transferring the artwork master to many printed circuit boards. Although silk fabric was originally used, today's screens are made of a very fine wire mesh, nylon, or polyester material. The screen is a very fine grid work of fibers. A frame is used to hold the screen at a constant tension.

The method described is a direct screen method. The indirect screen method is very similar.

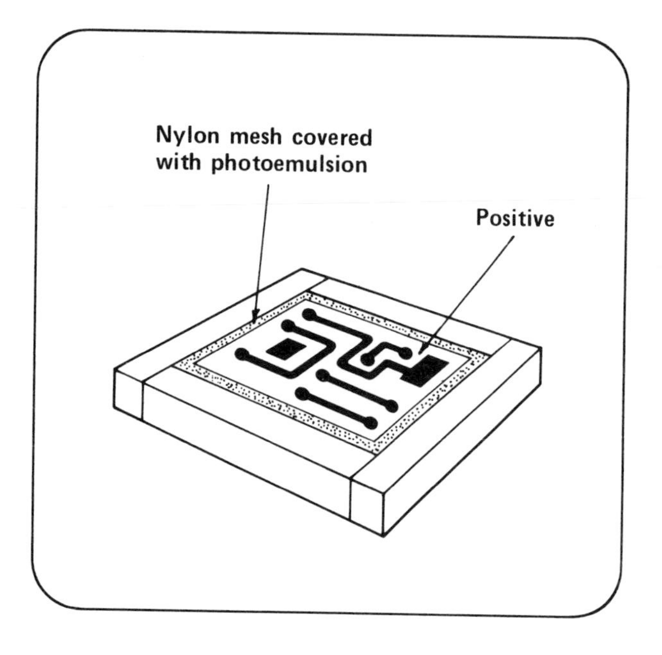

The screen is covered with a photographic emulsion and a photographic positive of the master drawing is placed on the screen. The photographic positive is sometimes referred to as a transparency. It is a copy of the original artwork master at the correct size for the PCB. The photographic positive has black lines for the conductor patterns and everything else is transparent.

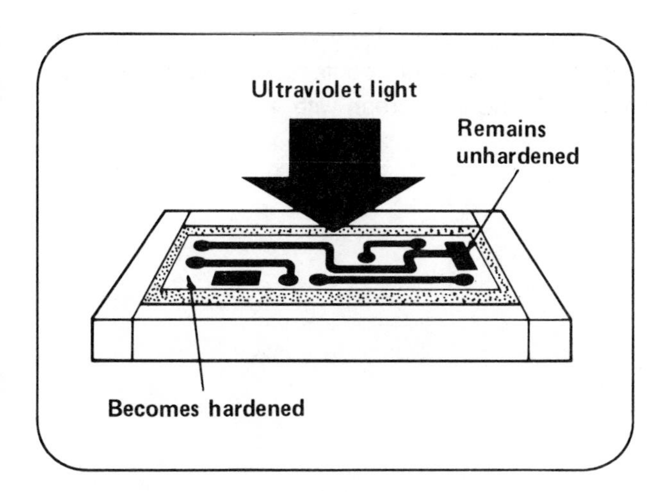

The screen is then exposed to ultraviolet light which "hardens" the emulsion that is not covered by the conductor patterns on the photographic positive that is exposed to the light. The emulsion lying underneath the conductor pattern of the photographic positive is not exposed and remains unhardened.

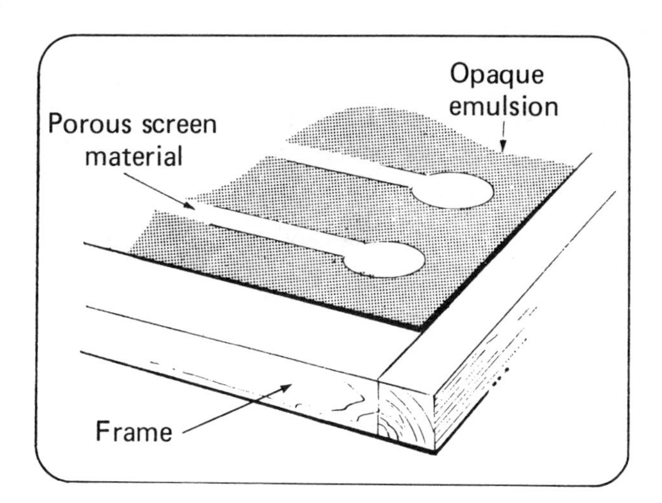

Developer is applied to wash away the unhardened emulsion leaving the hardened emulsion still in position. The original porous screen material in the shape of the conductor pattern is now surrounded by an opaque area of hardened emulsion.

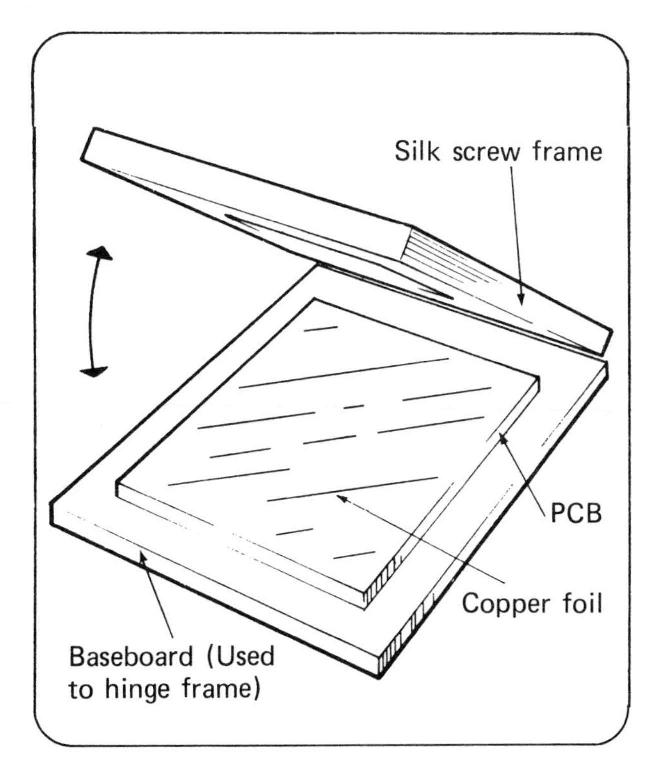

Using the silk screen

The unetched printed circuit board is positioned under the silk screen with the copper foil side toward the silk screen.

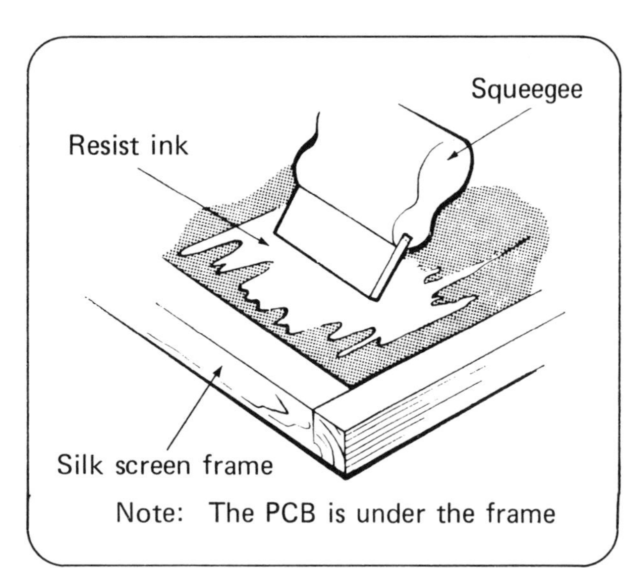

The frame is rotated down until the screen and the copper foil come into contact. A resist ink is then applied to the silk screen with a squeegee.

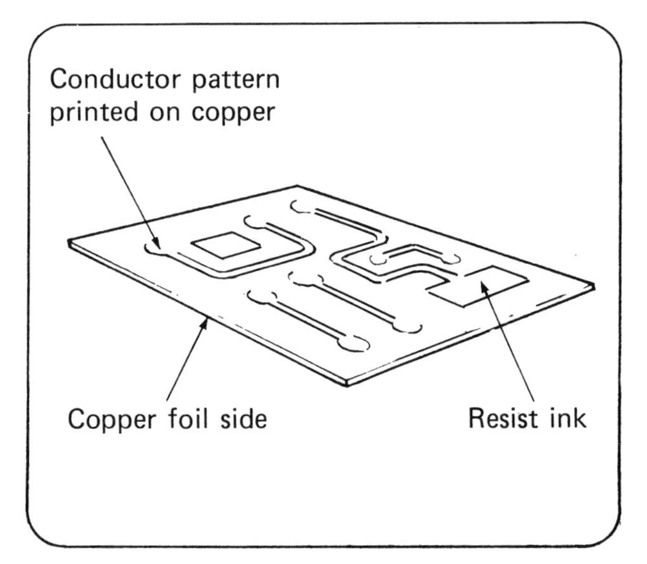

The resist ink is pushed or squeezed through the porous parts of the screen and printed onto the copper foil side of the circuit board.

The ink will resist the effect of the etchant.

The board is removed from under the silk screen so that another board may be inserted and printed.

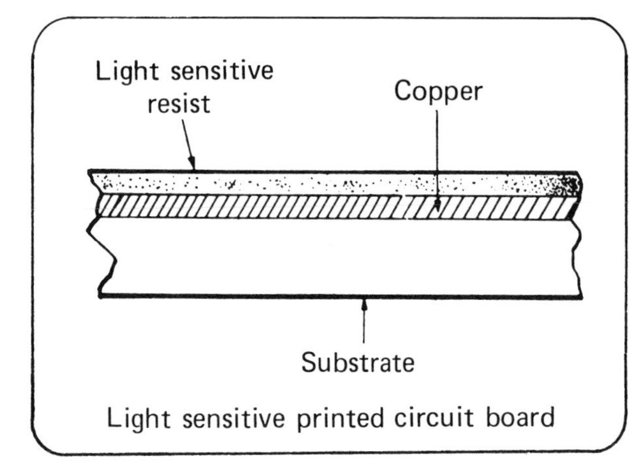

Light sensitive printed circuit board

The photographic method

The copper side of the printed circuit board is sprayed with or dipped in a light sensitive solution or resist.

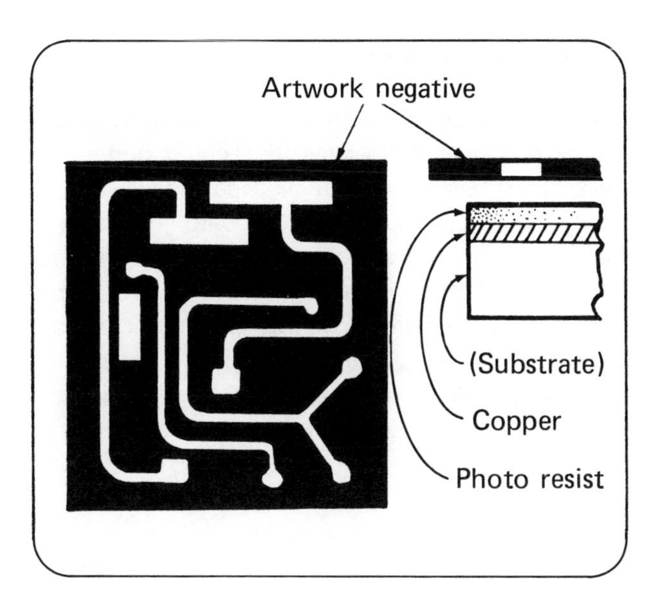

The photographic method normally requires a negative of the master artwork. The silk screen method required a positive.

The photonegative is contact printed on the copper side of the printed circuit board.

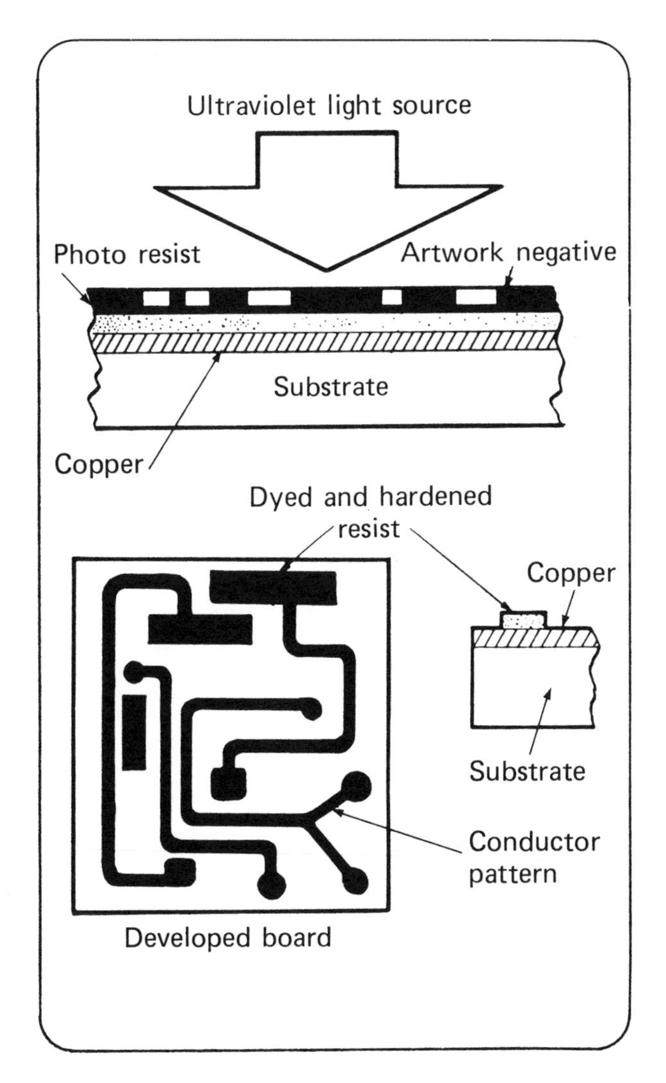

The board and the negative are exposed to ultraviolet light which hardens the resist along the conductor path where the light strikes. The rest remains unhardened.

The board is then placed in a developer solution that washes away the unhardened resist. The developer may contain a chemical dye so that the conductor paths are made easily visible. The board is now ready for etching.

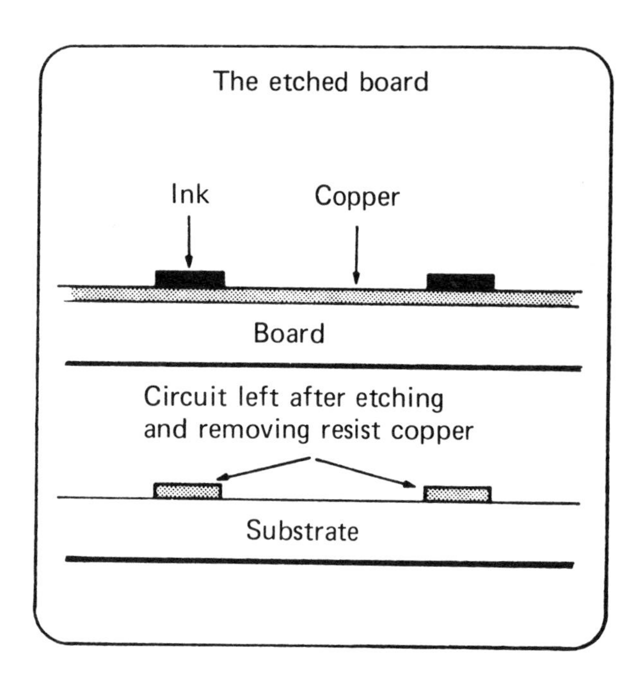

Etching the board

The board is sprayed with a chemical that removes the bare copper but leaves the copper under the resist. The resist is then removed leaving the printed circuit pattern, in copper, on the board. The chemical used to remove the copper is called an etching agent. The most common etchants are ferric chloride, ammonium persulfate, chromic acid, and cupric chloride. After etching, the board is thoroughly cleaned.

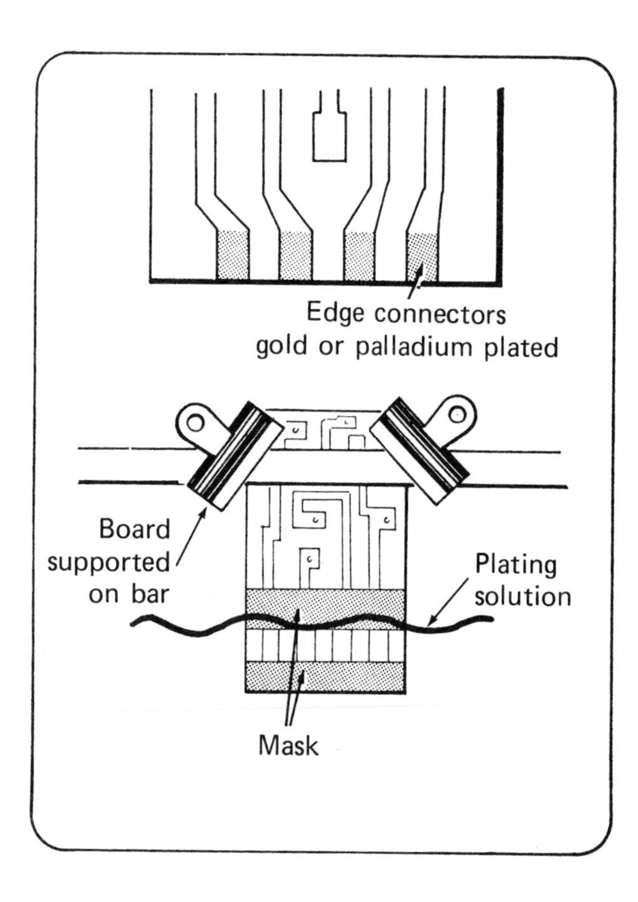

Edge connectors

Printed circuit boards are often required in rack mounts. The printed circuit boards are manufactured with edge connectors that match an edge connector socket. These edge connectors must be durable and allow little oxidation; these connectors are normally plated with gold or palladium. The board is masked (covered) so that only the copper required for the edge connectors is plated. The mask is removed after the plating operation.

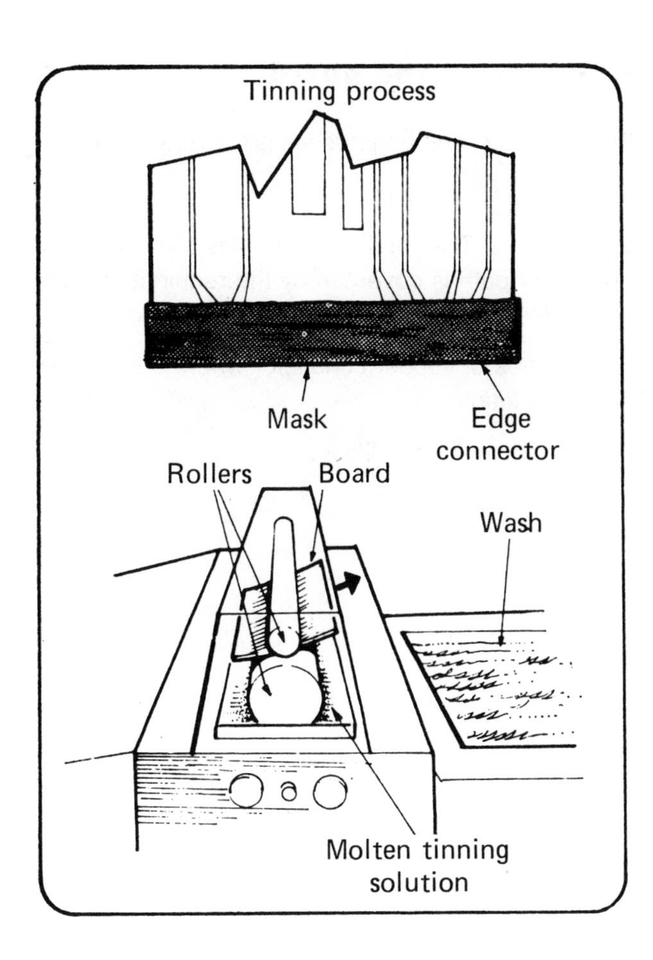

Tinning

After the board has been etched and washed, the conductor patterns may be tinned. Tinning helps to increase solderability and decrease the oxidation.

To complete the tinning, the manufacturer feeds the board between two rollers with the copper side down. The lower roller revolves in a bath of molten tinning solution. The conductor paths become covered with an even layer of tinning solution. The board is then washed.

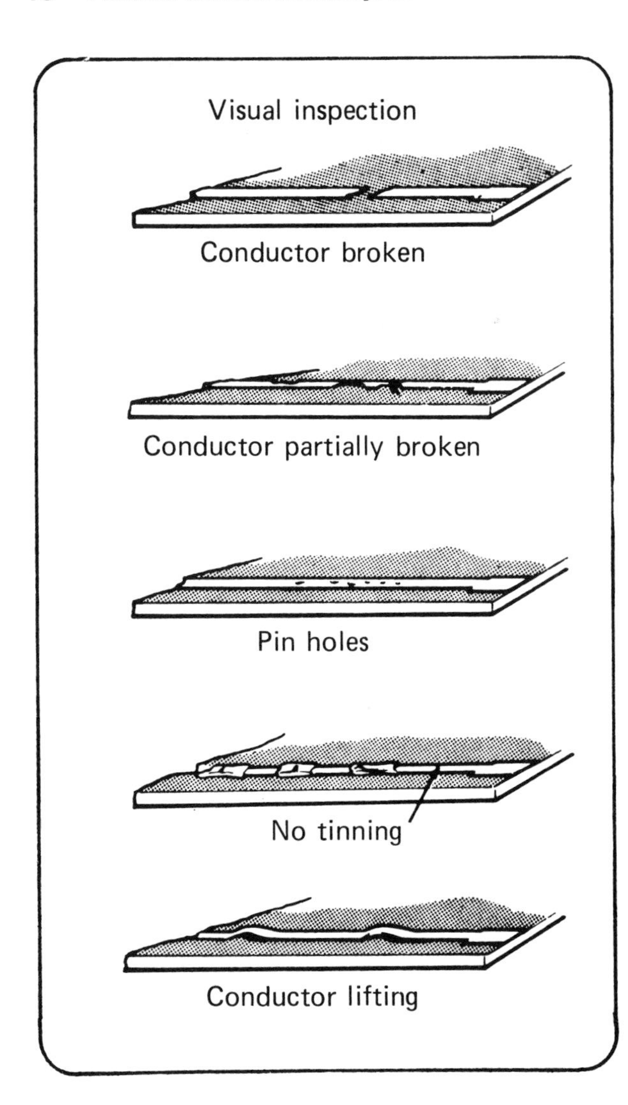

Completing the board

The printed circuit board is visually inspected for

1 Broken conductors.

2 Pinholes. The number of pinholes allowed in the conductor path is governed by the required specification. No pinholes are allowed in edge connectors.

3 Patches of copper conductor where there is no tinning solution.

4 Conductor or pad lifting.

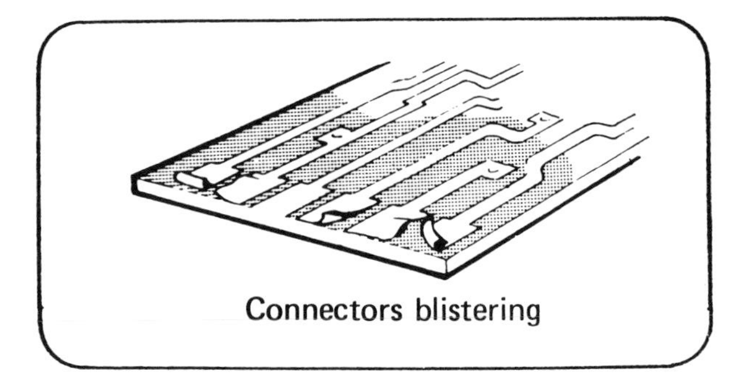

5 Blistering of edge connectors.

6 Correct positioning and diameters of holes (after drilling).

Printed circuit boards that have passed the visual inspection are normally rubber-stamped or identified in some way.

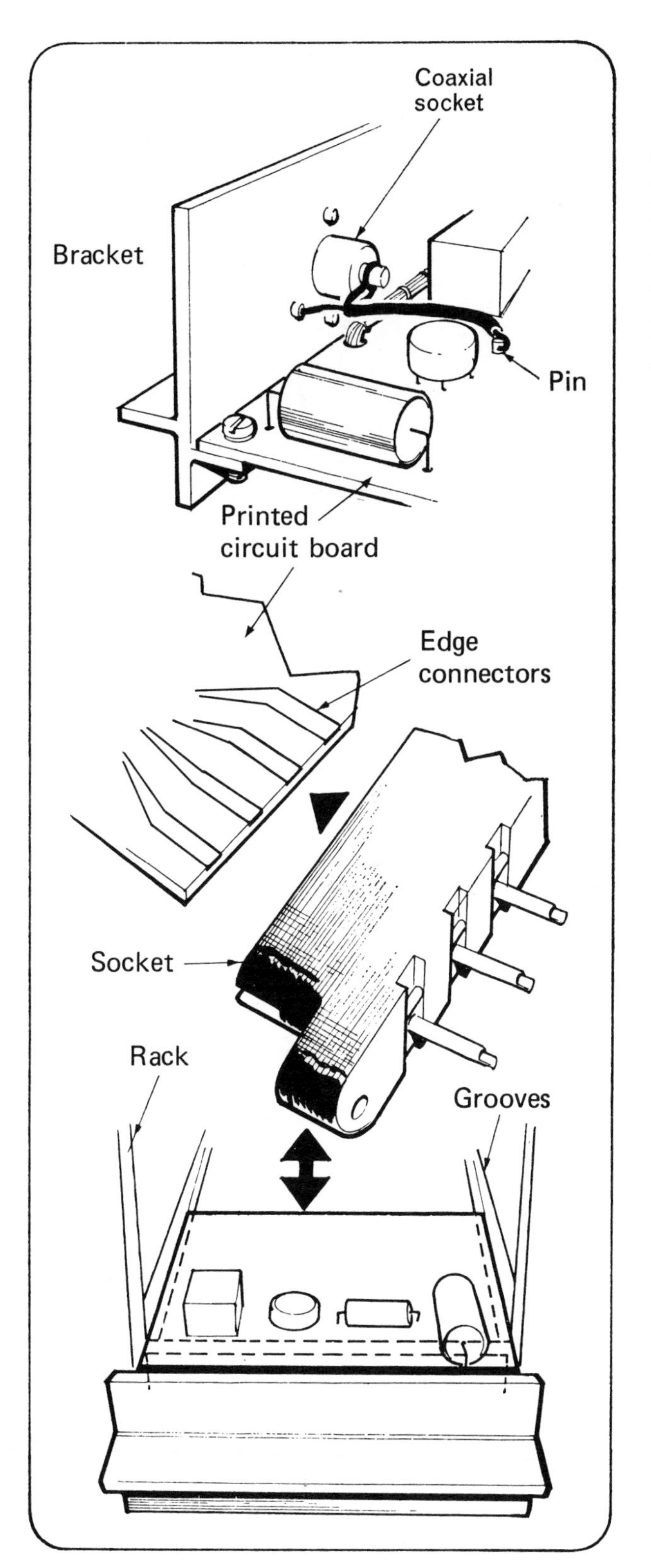

External connections to printed circuit boards

Connections to the board may be made by either pins or edge connectors. A typical printed circuit board is shown. The board is attached to a bracket that serves as a support. A connection is made from the coaxial socket to a pin on the printed circuit board.

The edge connector mates with a rack or panel mounted socket. The printed circuit card slides along grooves in the rack and can be quickly inserted or removed.

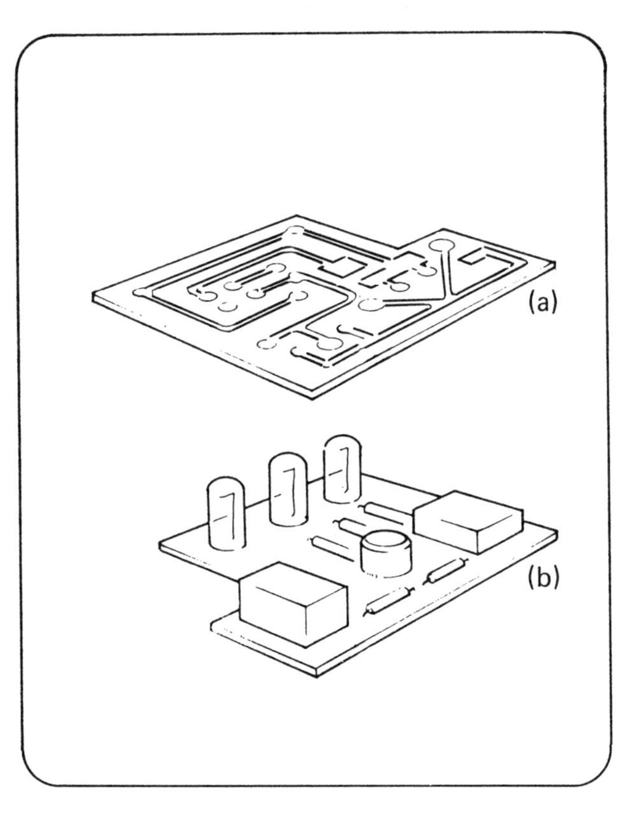

Test items

1 Label the component side and the conductor side of the printed circuit board.

____

(See page 2.)

2 The silk screen method requires a photographic

____

(See page 4.)

3 The photoemulsion used on a silk screen is sensitive to ______ light.

(a) red

(b) ultraviolet

(c) infrared

(See page 5.)

4 The photographic method of manufacturing printed circuit boards requires a photographic

____

(See page 7.)

5 Circle two chemicals commonly used for etching printed circuit boards.

(a) Ferric chloride

(b) Sodium chloride

(c) Ammonium persulfate

(d) Hydrochloric acid

(See page 8.)

Module two

Printed circuit board components

Introduction

This module shows how electronic components are mounted on printed circuit boards. New component configurations have been developed to meet the special demands of printed circuit technology. Some of these components are shown.

Key technical words

- **Heat sink** A device used to remove heat generated by a transistor.
- **Transistor socket** A device that is soldered to the PCB, allowing a transistor to be inserted.

Performance objectives

Upon completion of this module, you should be able to:

- State the advantages and disadvantages of a flush mounted component.
- State why a heat sink is used.
- Describe mounting techniques for transistors, resistors, capacitors, relays, tubes, and diodes.

Components for printed circuit boards

Resistors

Specialized printed circuit resistors are shown to the left.

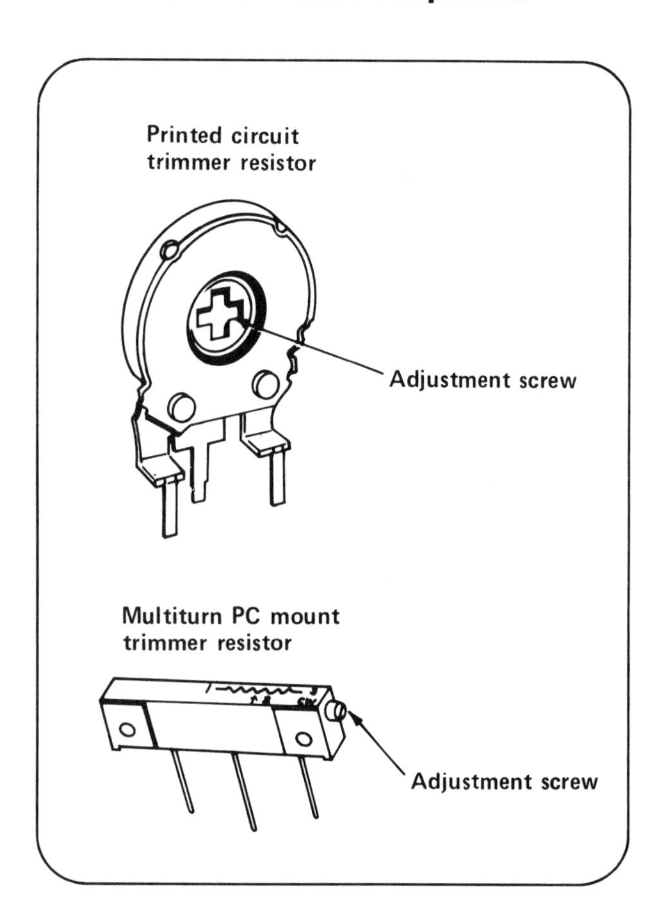

The resistors normally mount flush with the board. This provides mechanical strength to the component. Vertical mounting is used to save space or where density packaging is required.

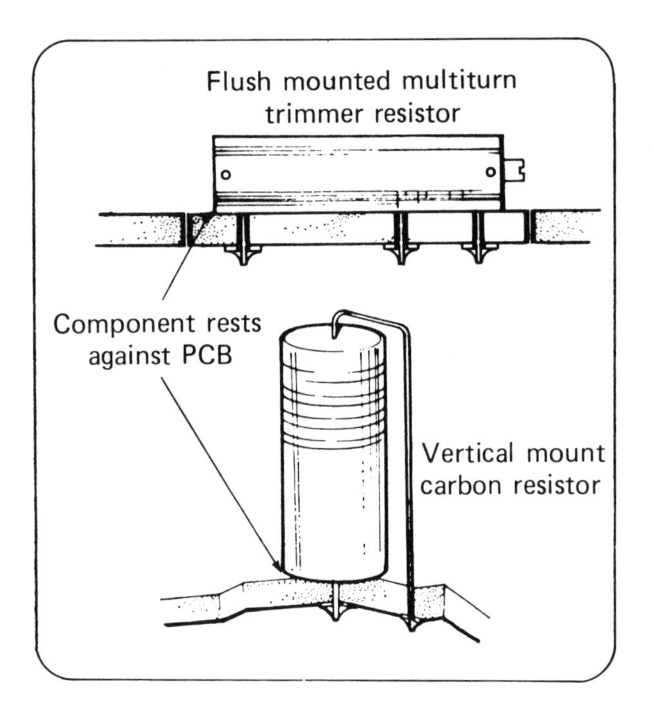

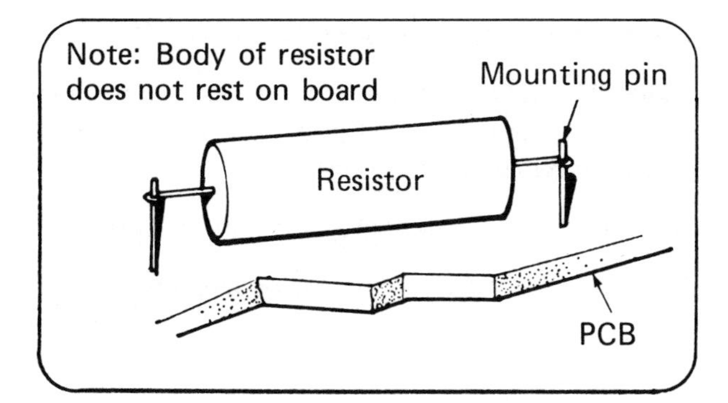

A high wattage resistor may need additional cooling. The resistor is mounted above the board. The space between the resistor and the PCB will increase air flow and help cool the device. The resistor is rigidly supported.

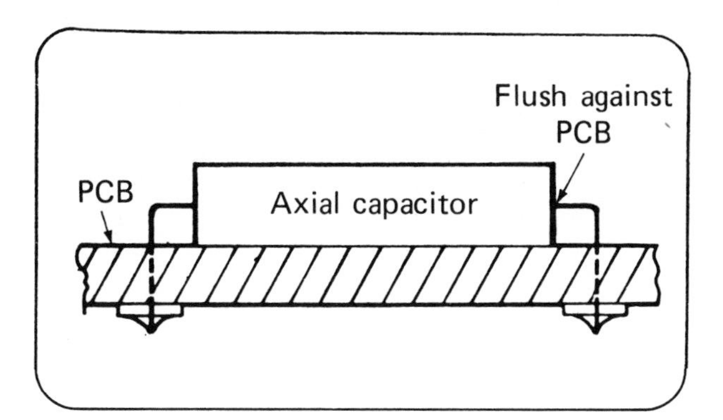

Capacitors

Axial capacitors are mounted flush with the board. Solid state *diodes* are also mounted using this method.

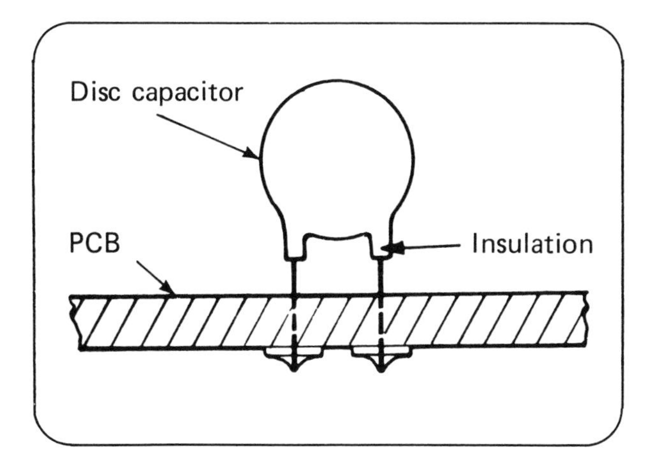

Disc type capacitors

The body is held above the board to insure that the insulation does not break or melt.

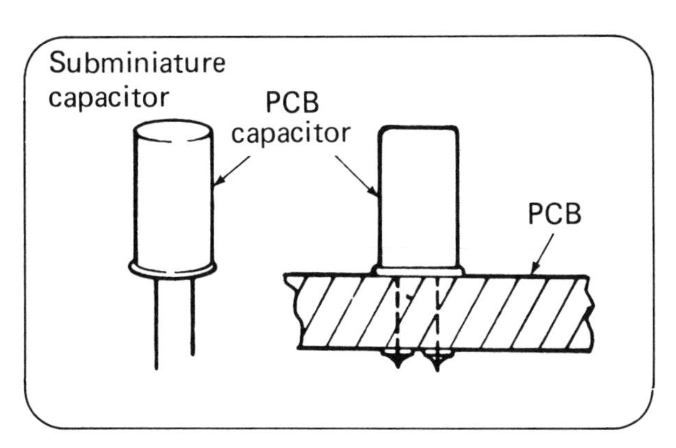

Specialized capacitors designed for use on printed circuit boards mount flush with the board.

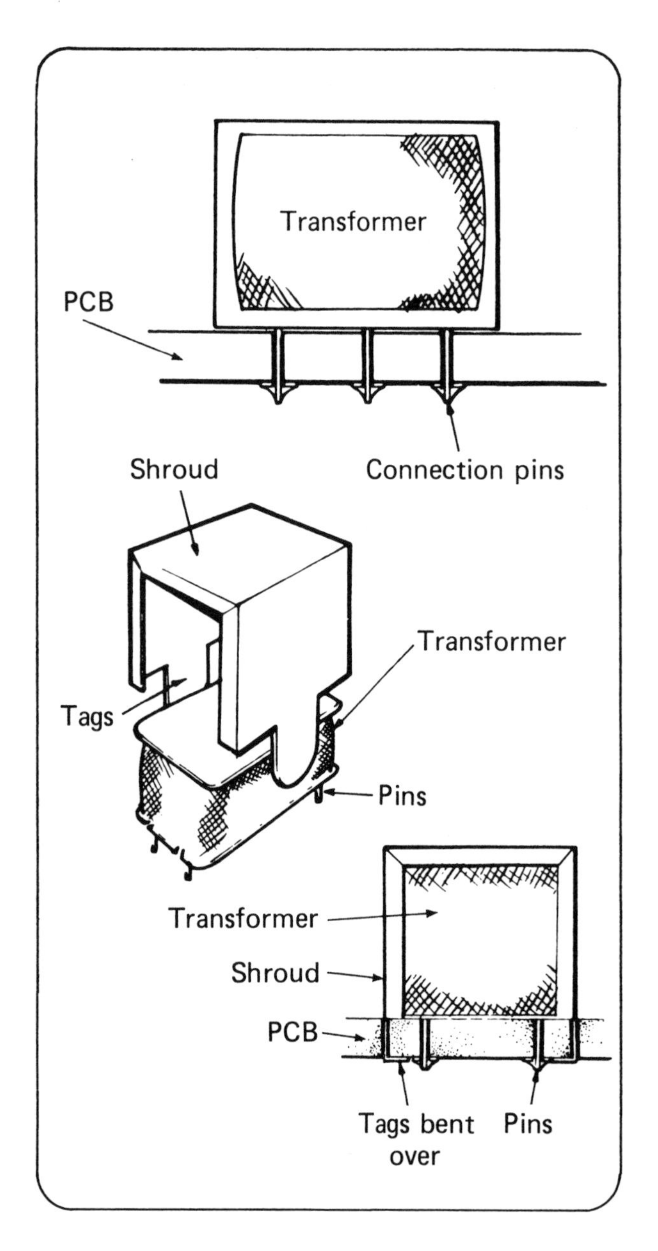

Transformers—PCB mount

Although many transformers are too large and heavy to mount on a PCB, miniature transformers may be mounted on the board.

The transformer's connection pins are soldered.

The shroud tags are bent over to secure the shroud.

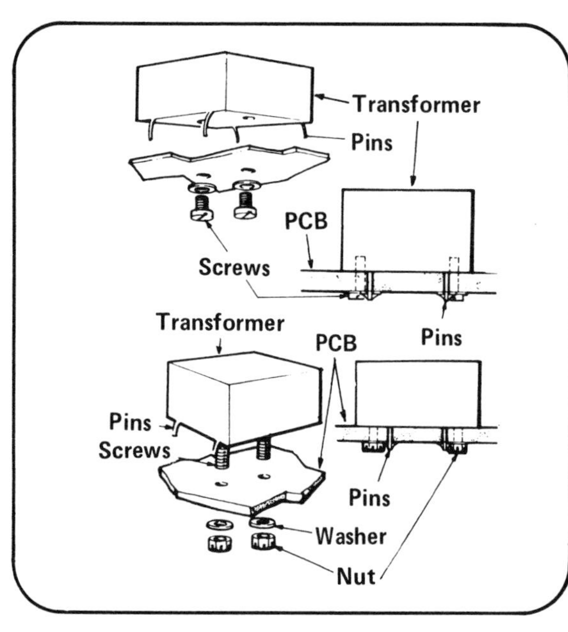

Encapsulated transformers are usually provided with mounting screws or lugs that are used to mount the transformer to the board.

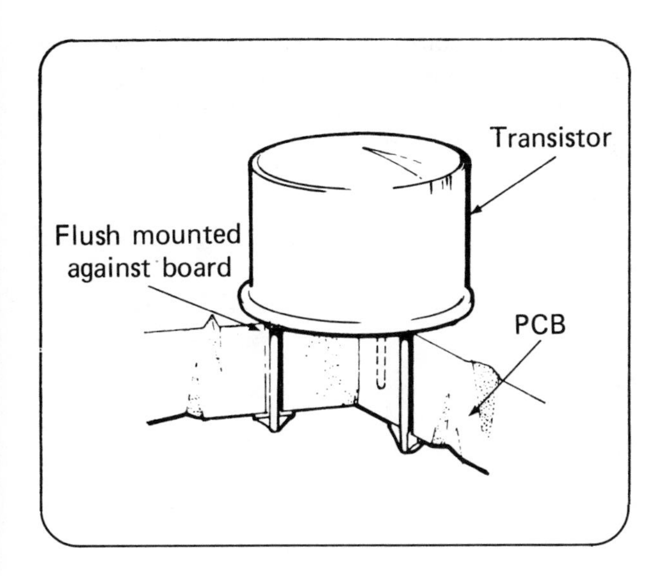

Transistors

Transistors are mounted in a variety of ways on printed circuit boards.

1 The transistor is mounted flush with the board. This could cause heating problems, but it does provide mechanical security for the transistor.

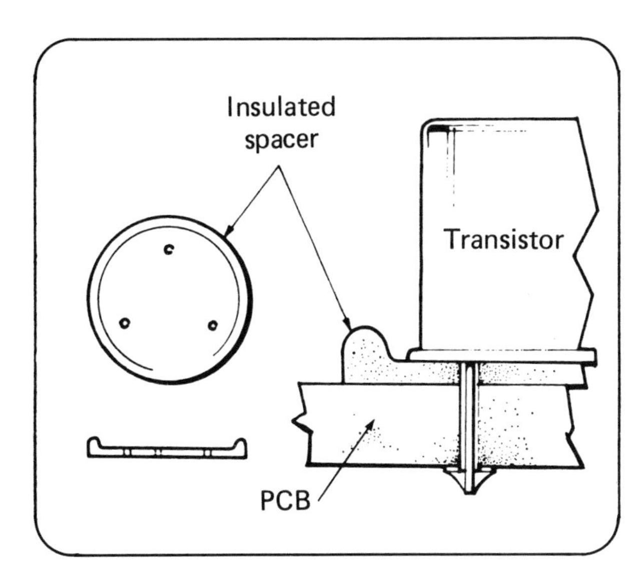

2 The transistor body is spaced above the board by an insulated spacer. The spacer provides stress relief for the transistor leads.

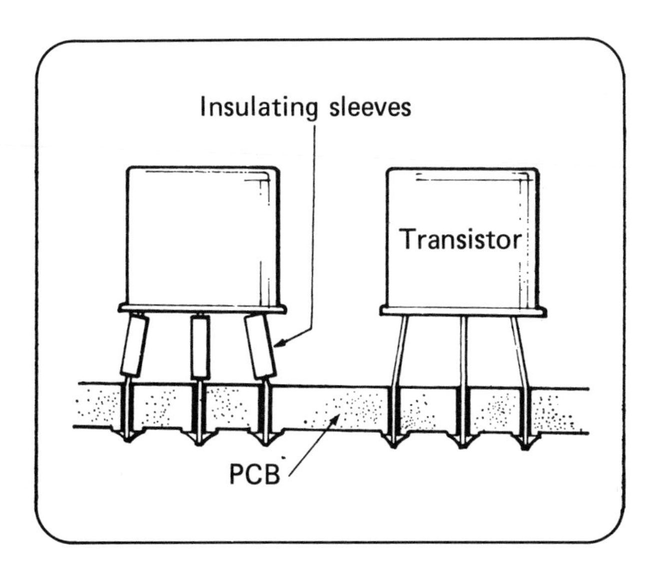

3 The transistor body is mounted above the board. This provides less mechanical strength, but better heat dissipation. Insulating sleeves may be placed around the leads.

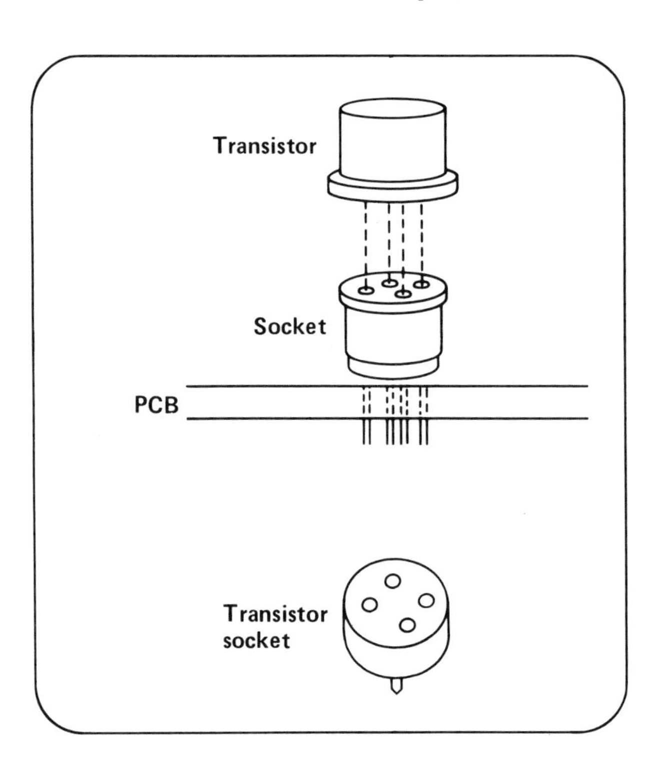

4 Transistor sockets are soldered to the PCB. The transistor is then inserted into the socket. This mounting is very helpful in trouble-shooting and repair.

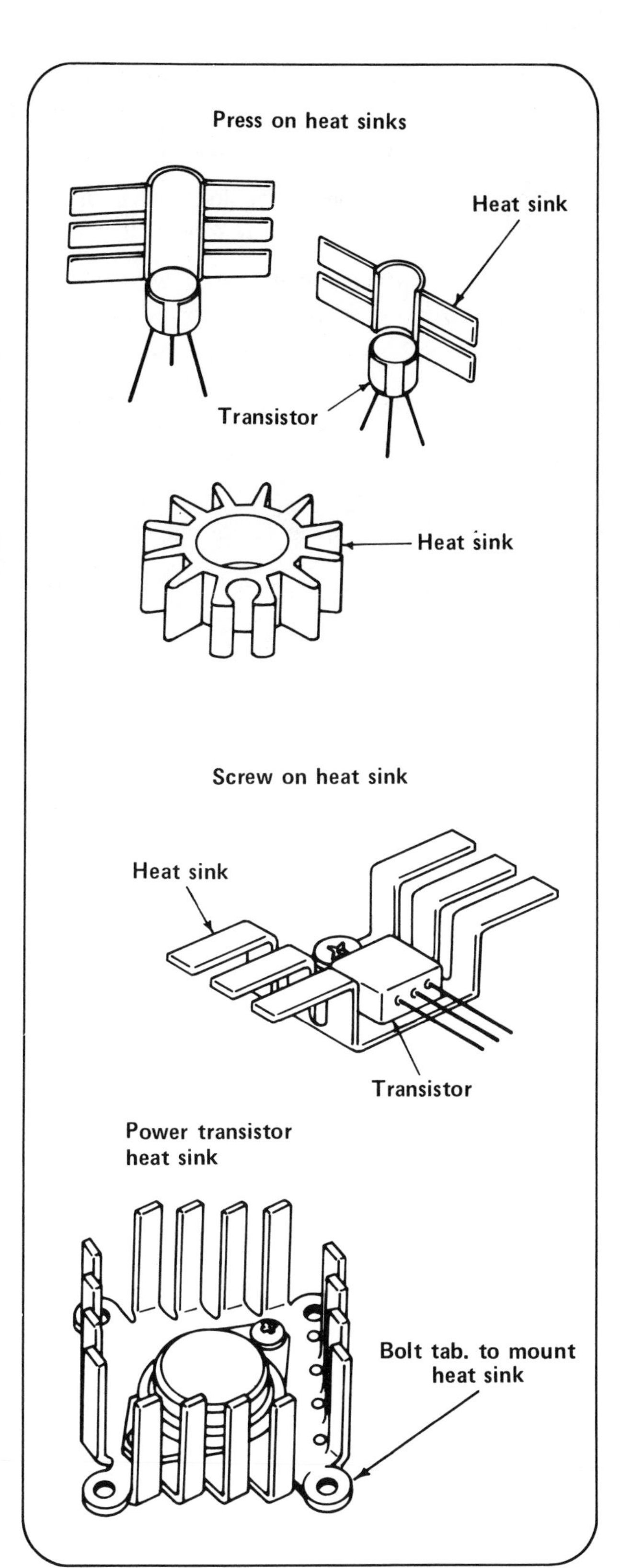

Heat sinks are used with transistors. The heat sink is used to dissipate the heat generated by the transistor. There are many styles of heat sinks.

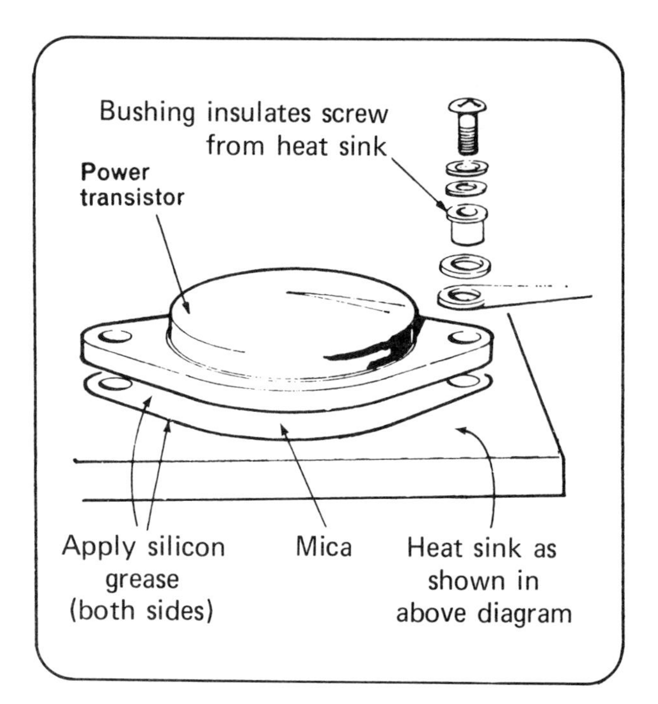

To mount a power transistor on a heat sink

1 Clean the heat sink surface and remove any burrs around the holes.

2 If electrical isolation is required between the transistor and the heat sink, use a mica insulator.

3 Coat both surfaces of the mica with silicone grease to provide good heat transfer between the transistor and the heat sink.

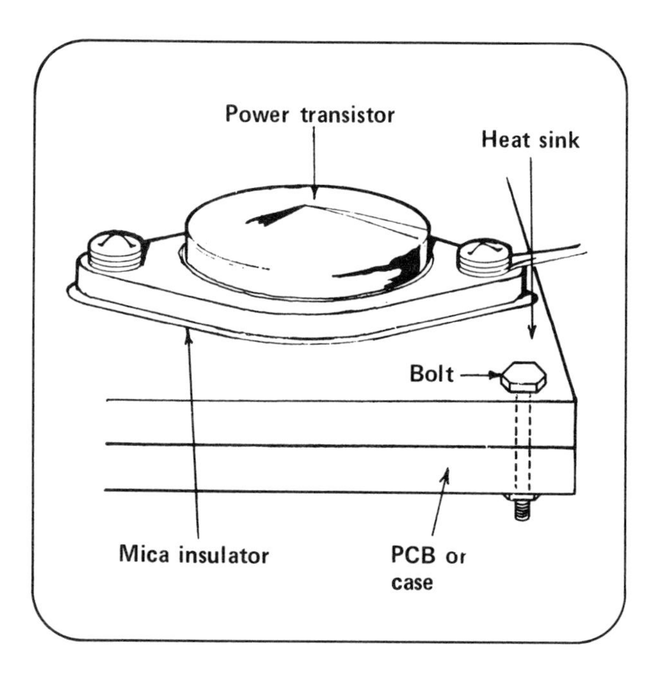

4 The heat sink is usually mounted to the PCB or case with bolts. The leads of the transistor are soldered to the conducting pads.

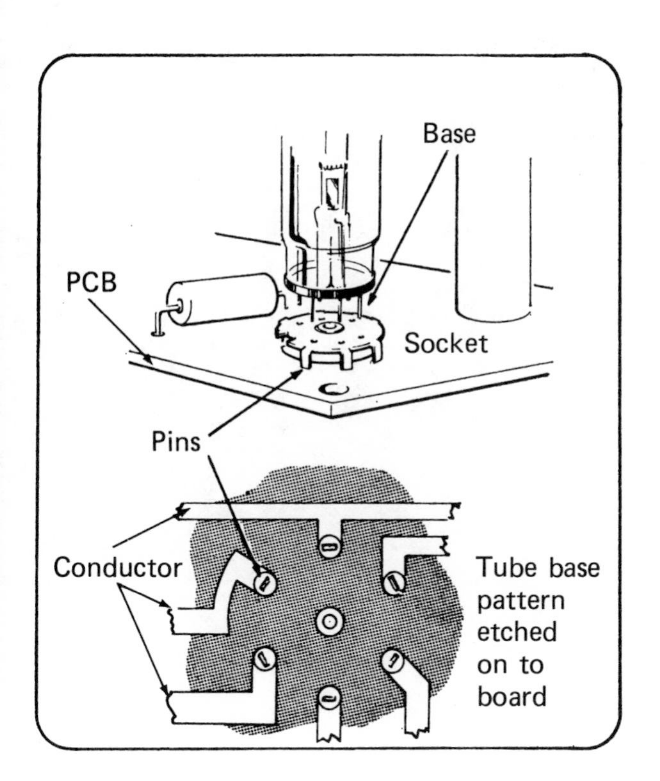

Tube sockets

Printed circuit tube sockets are mounted on the board by passing the socket pins through holes in the board and soldering them to the pads.

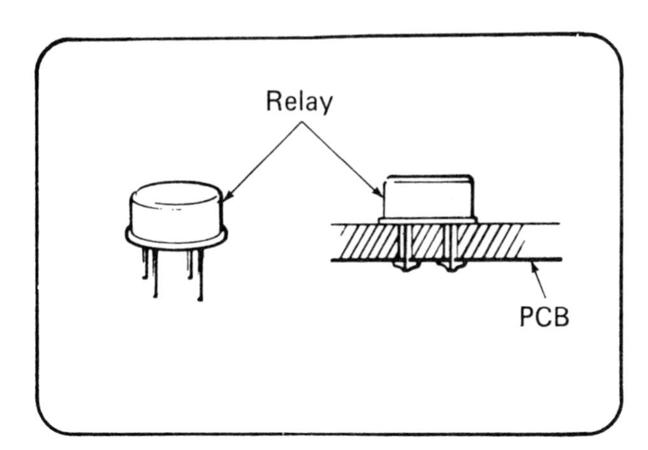

Relays

Special printed circuit relays mount flush with the board.

Test items

1 Which has better mechanical strength?

(a) A flush mounted transistor

(b) A transistor mounted above the board.

(See page 17.)

2 Silicon grease is applied to _______ sides of the mica insulator?

(a) only one of the

(b) both

(See page 20.)

3 Which type of mounting requires less board space for an axial resistor?

(a) Vertical mounting

(b) Horizontal mounting

(See page 14.)

Module three

Printed circuit board assembly

Introduction

This module deals with how the components and the board fit together to make the finished PCB. Soldering fixtures such as vises and lead benders are discussed in this module.

Various sets of drawings are used to show the order of the component mounting and the position in which they are mounted on the board. Handling precautions, the tools necessary to assemble a board, and the final inspection process are covered in this module.

Key technical words

- **Resin core solder** A type of solder that contains flux already in the solder.
- **Lead bender** A specialized tool used to preform component leads before they are inserted in the PCB.
- **Desoldering** The process of removing solder from a solder joint.

Performance objectives

Upon completion of this module, you should be able to:

- Identify a faulty conductor path or pad.
- Describe when insulated leads are required on a component.
- Explain when a lead heat sink should be used.

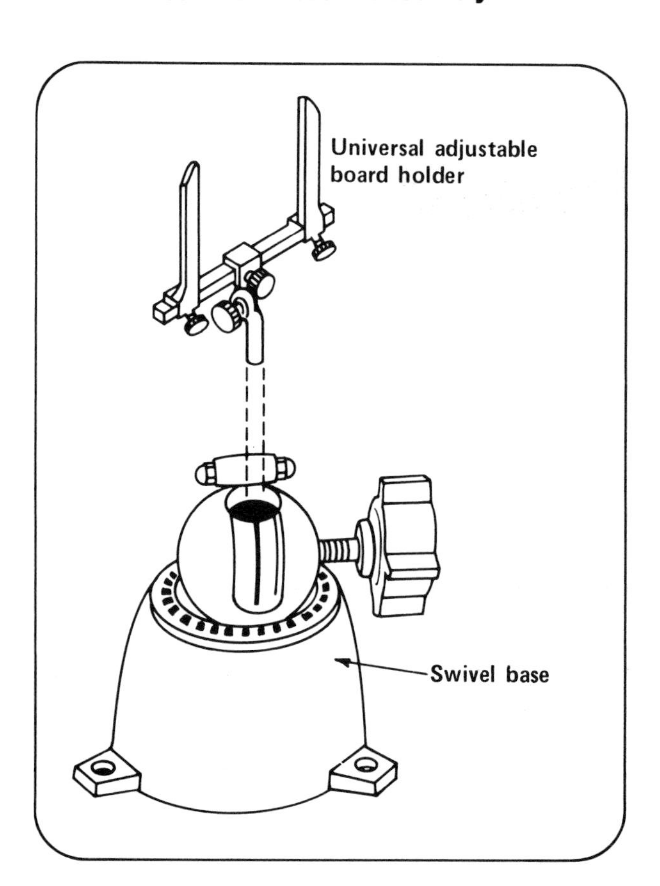

Printed circuit board assembly

Soldering fixtures

There are several types of printed circuit board vises and board holders available. The vise holds the board while components are mounted and soldered.

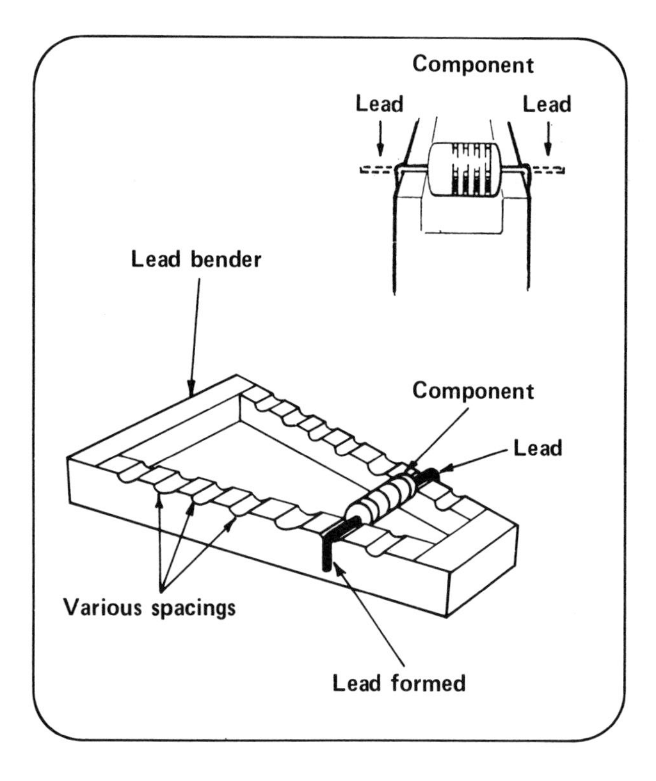

Preforming jig

This device is also called a lead bender or bending block. It is used to shape component leads. The jig has various widths that correspond to the distance between lead holes on the PCB.

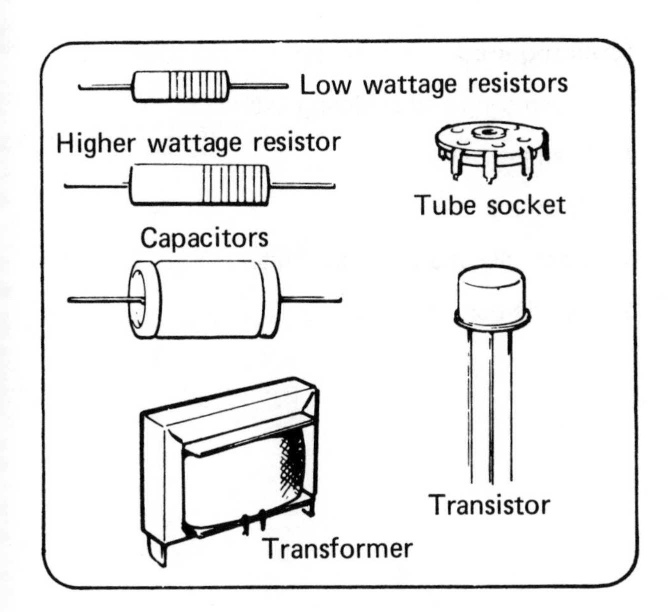

Soldering components to a printed circuit board

The following components are soldered directly to PCBs.

Low wattage resistors.
Tube sockets.
Capacitors.
High wattage resistors.
Transistors and transistor sockets.
Transformers.

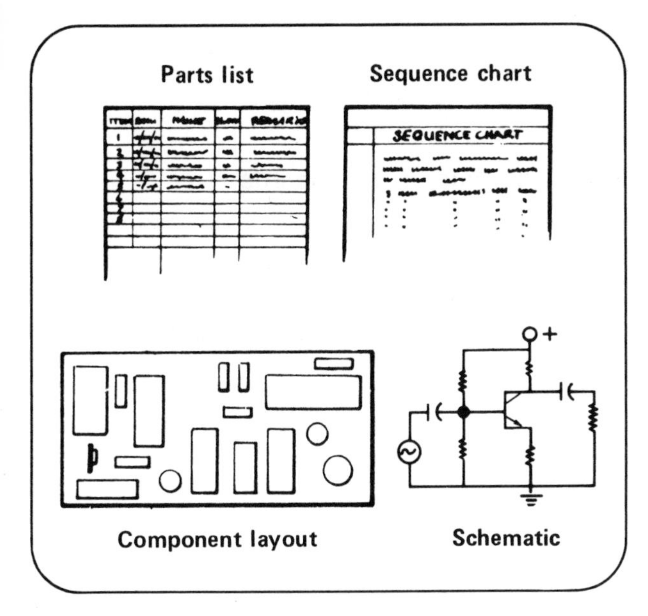

Drawings

Drawings consist of

Parts list A complete list of all parts. Each part has an identification number.

Component layout Shows placement of each part on the PCB.

Sequence chart Lists the order the parts are mounted on the PCB.

Schematic An electronic diagram of the circuit. (Not always provided in assembly drawings.)

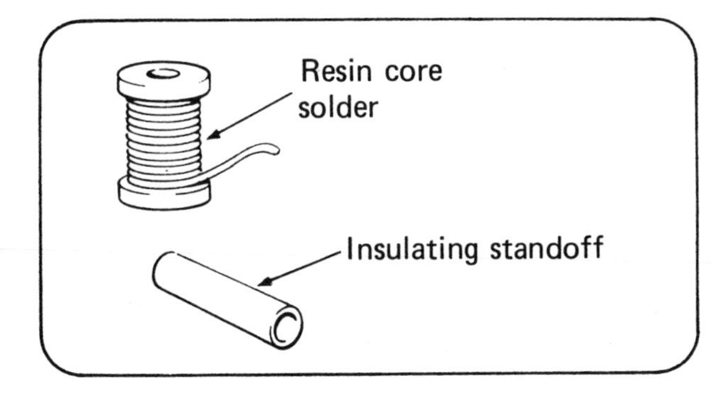

Materials required

1 Resin core solder.

2 Insulating standoffs.

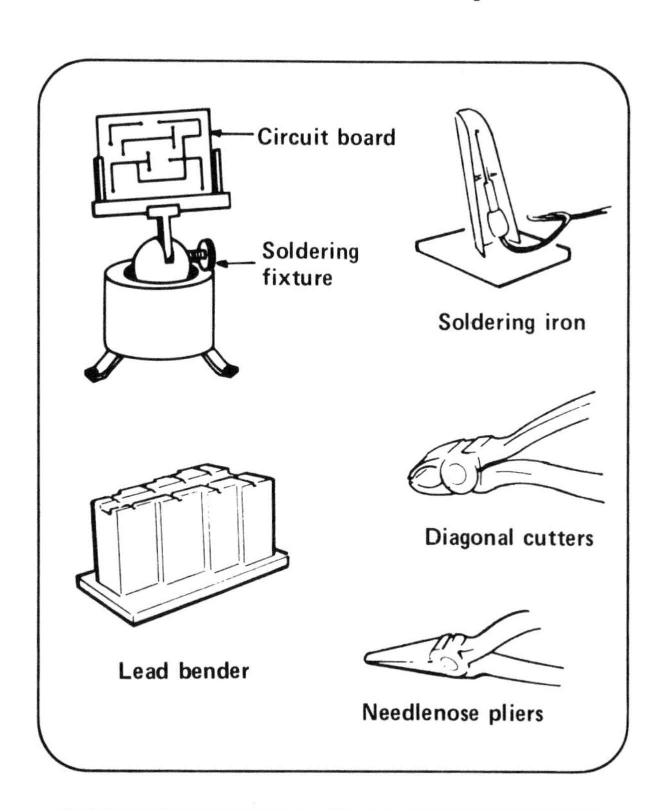

Tools required

1 Assembling and soldering fixture.

2 Lead bender.

3 Soldering iron.

4 Diagonal cutters or flush cutters.

5 Needle-nose pliers.

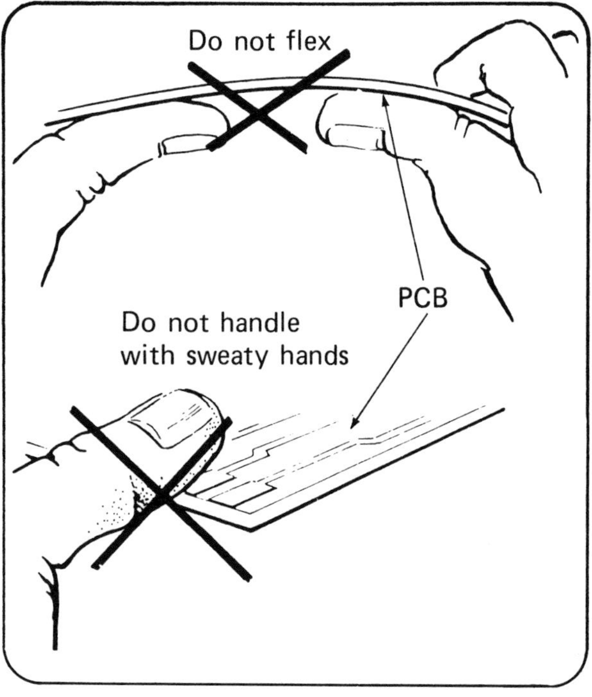

Handling printed circuit boards

Do not flex a board because this could cause breakage across conductors. Do not handle a board with sweaty hands. This may cause the conductors to corrode. Handle the boards by their edges.

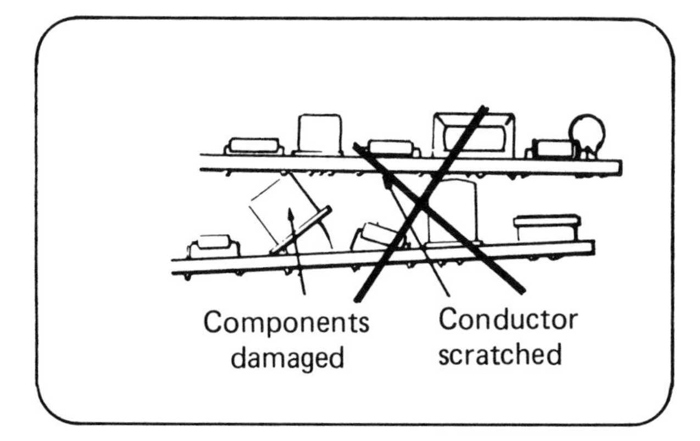

When handling printed circuit boards on which components have been mounted

1 Do not stack one board on top of another because conductors may become scratched and components damaged.

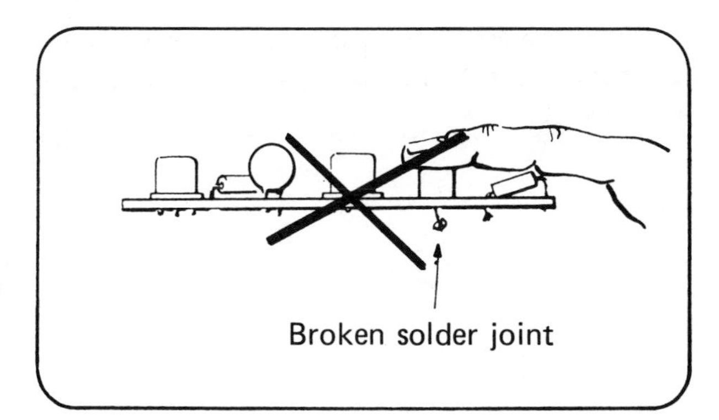

2 Do not apply pressure to components because this can damage the solder joint.

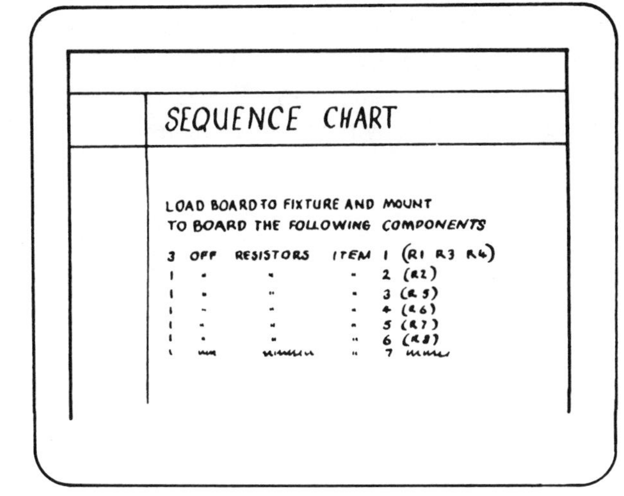

Mounting the low wattage resistors and the tube socket

1 Using the parts list, determine the required components.

2 Refer to the sequence chart to determine the order in which the components should be mounted. The chart indicates that resistors R1, R3, and R4 should be mounted first.

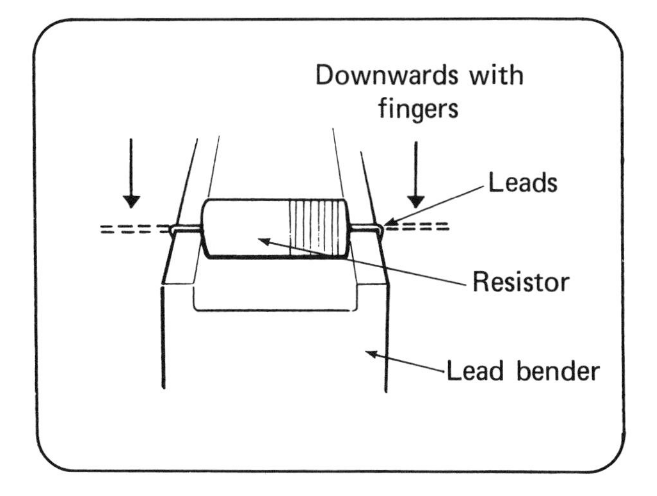

3 Take each of the resistors and place it over the correct part of the lead bender. Hold the resistor down and form the leads over the lead bender. This bends the leads to the desired right angle.

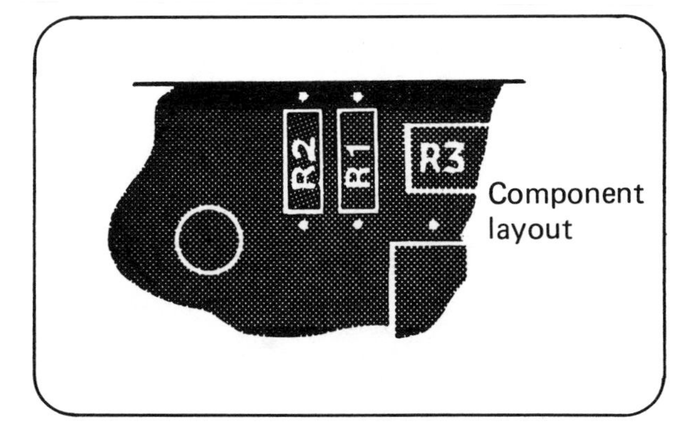

4 Refer to the component layout drawing and note the positions of resistors R1, R3, and R4.

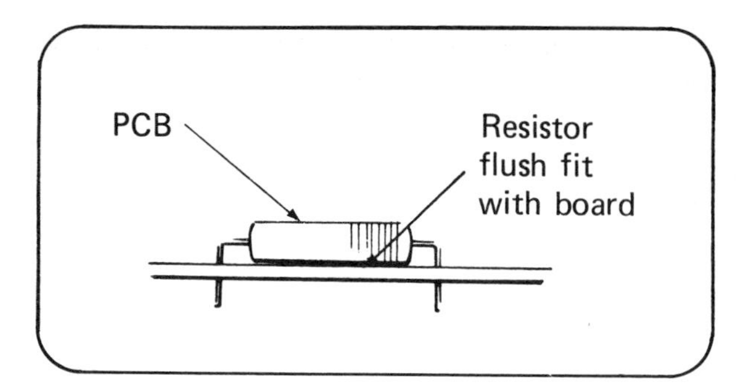

5 Place the leads of R1 through the holes in the board and push the resistor down so that it is flush with the board.

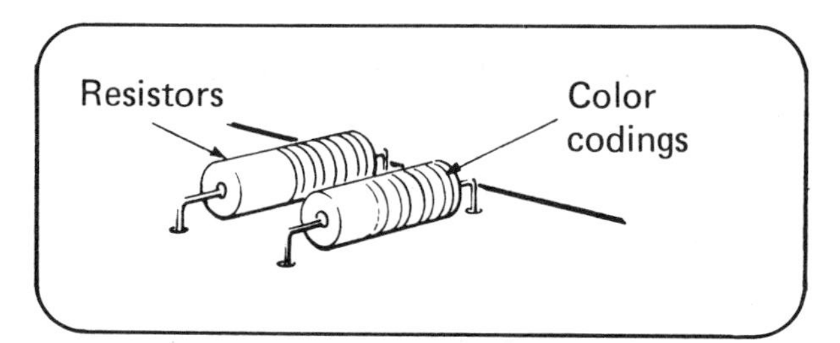

6 Repeat for resistors R3 and R4, and then for the rest of the resistors, according to the sequence chart. Be sure that the color coding on all parallel resistors runs the same way. This is required for appearance and easy part identification.

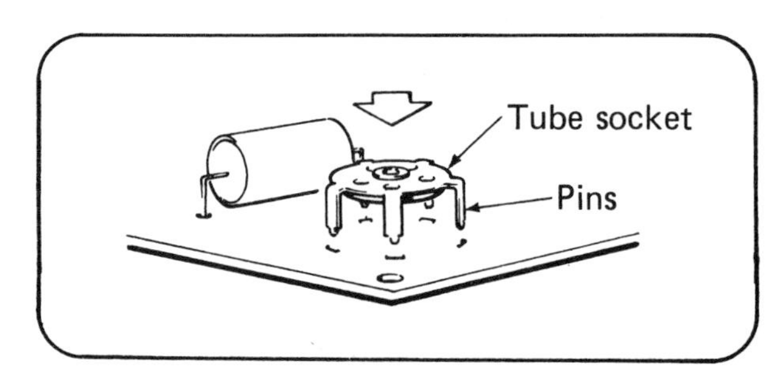

7 Take the tube socket, the next component on the sequence chart, and push the pins into the holes on the board.

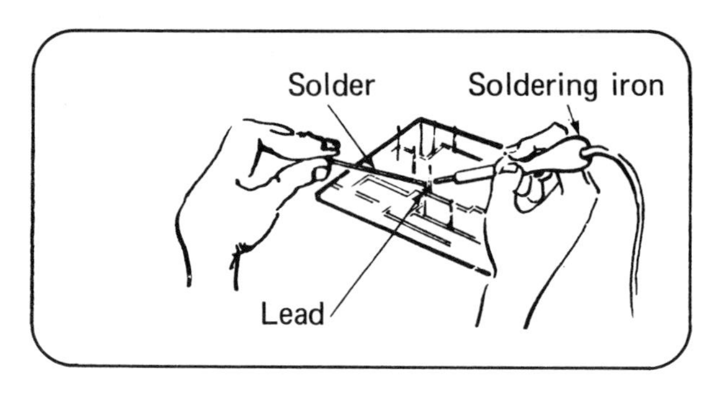

8 Solder each lead to its pad by touching the soldering iron and then the solder to the lead and pad. It will be necessary to support the components in the PCB card holder when you rotate the board to solder the components.

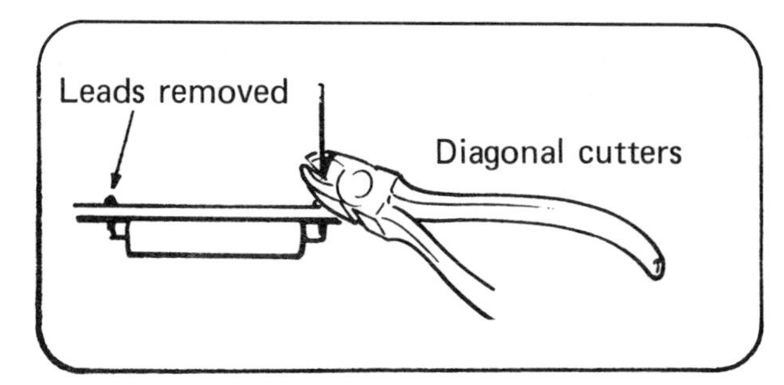

9 Cut off the ends of the component leads using the diagonal cutters.

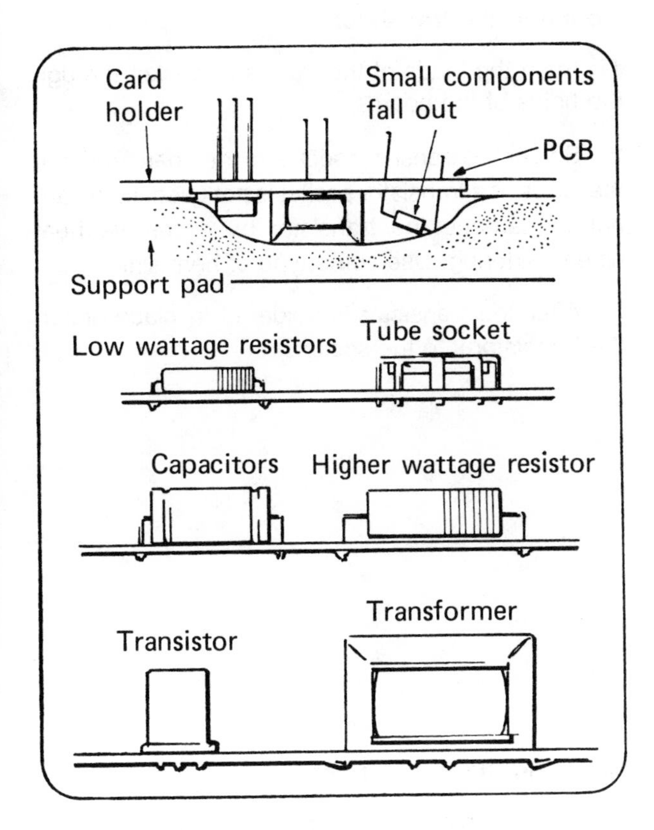

Order of component mounting

Components are mounted in the order shown on the sequence chart. The low wattage resistors and the tube socket have been soldered in position. These components have to be mounted on the board first because they are smallest.

If the larger components are mounted first, the smaller components will fall out when reversing the fixture.

Therefore, the smallest components are mounted first, then the slightly larger sizes, and so on until the largest components are in position.

In this exercise the order of size is

1 The low wattage resistors and tube socket.

2 The capacitors and the higher wattage resistor.

3 The transistor.

4 The transformer.

The capacitors and the higher wattage resistor are mounted next.

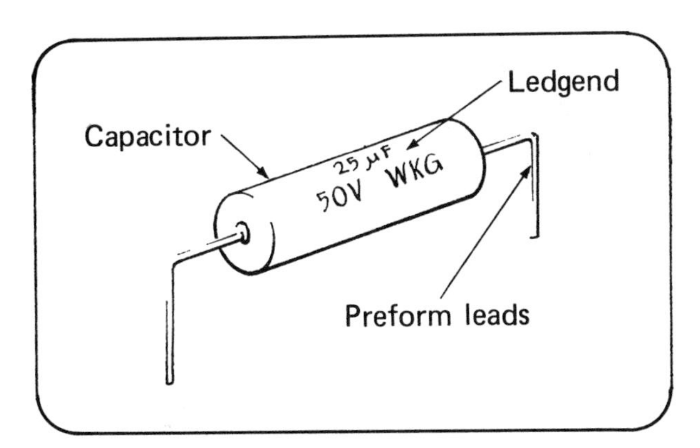

Mounting the capacitors and the high wattage resistors

1 Select the capacitors as indicated on the sequence chart.
Bend the leads and assemble them to the board.
Make sure that the legend showing capacitance, voltage, etc., is readable.
The legends on parallel nonelectrolytic capacitors should all read in the same direction.
Electrolytic capacitors are connected observing polarities.

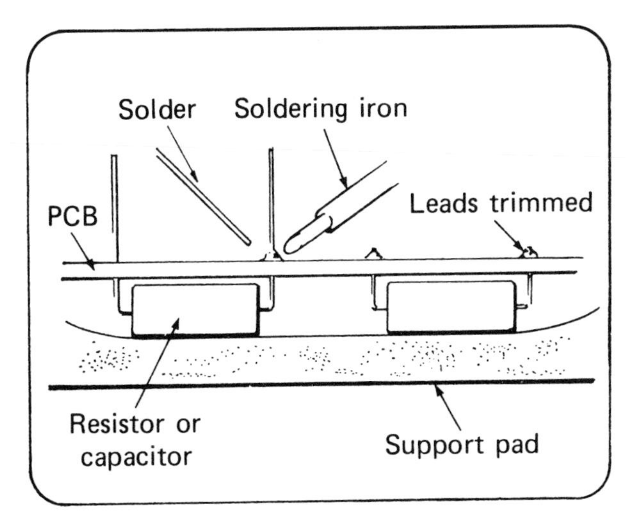

2 Bend the leads of the high wattage resistors as required.

3 Assemble the resistors onto the board. High wattage resistors are often mounted slightly off the board to aid in cooling.

4 Solder and trim the leads.

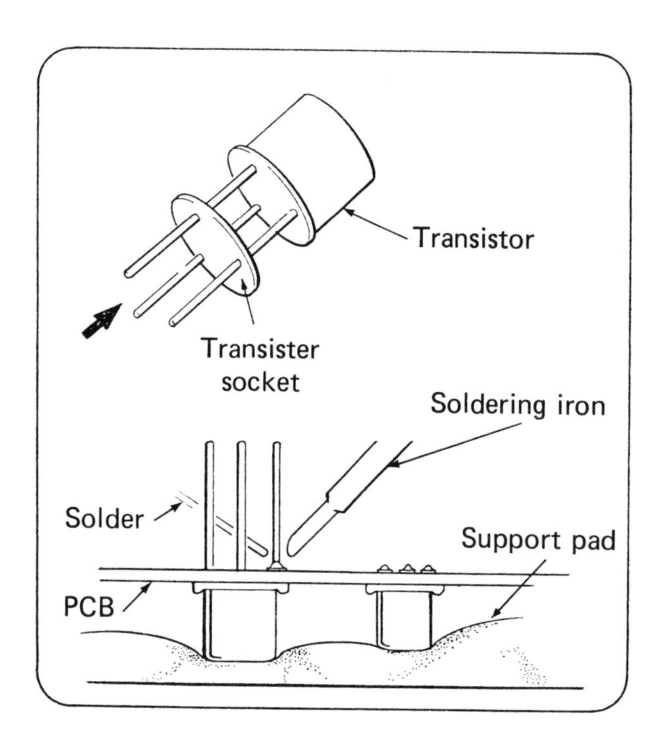

Mounting the transistor

1 Push the leads of the transistor socket through the holes of the board.

2 Solder the transistor socket to the board and trim the leads. If a transistor socket is not used, take care not to damage the transistor by excessive heat when soldering. Use some type of heat sink.

3 After the transistor is soldered in place, insert the transistor into the socket.

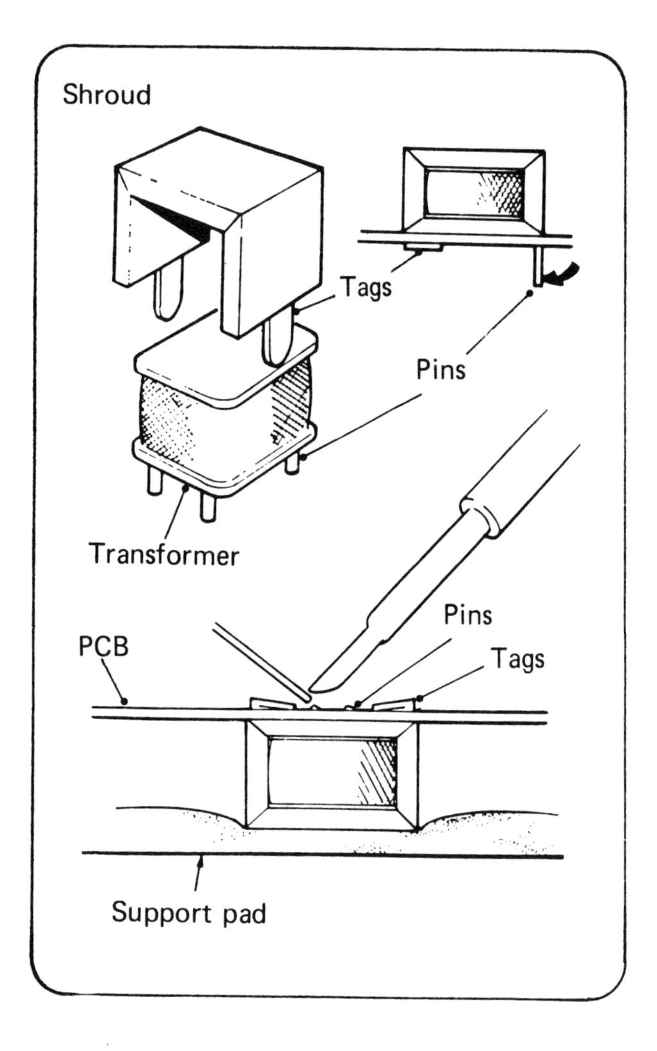

Mounting the transformer

1 Push the pins of the transformer into the appropriate holes in the board.

2 Push the shroud tags through the slots and bend them over.

3 Solder and trim the leads.

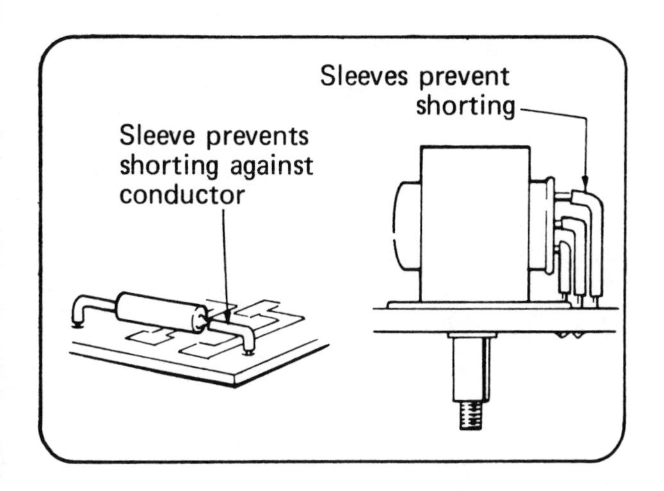

Assembly precautions

1 Insulated sleeves should be used on component leads when there is a possibility of short-circuiting other leads.
Double-sided boards may require more component sleeving.

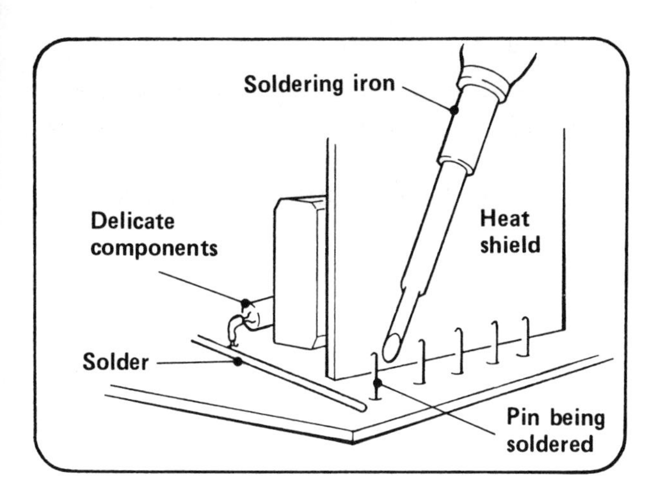

2 Heat shields or lead heat sinks are sometimes necessary to protect delicate components from heat caused by soldering. These devices are normally made from aluminum, copper, or brass which are all good heat conductors. The heat shield is placed between the soldering iron and nearby components. The heat sink is placed on the lead between the devices and the soldering iron.

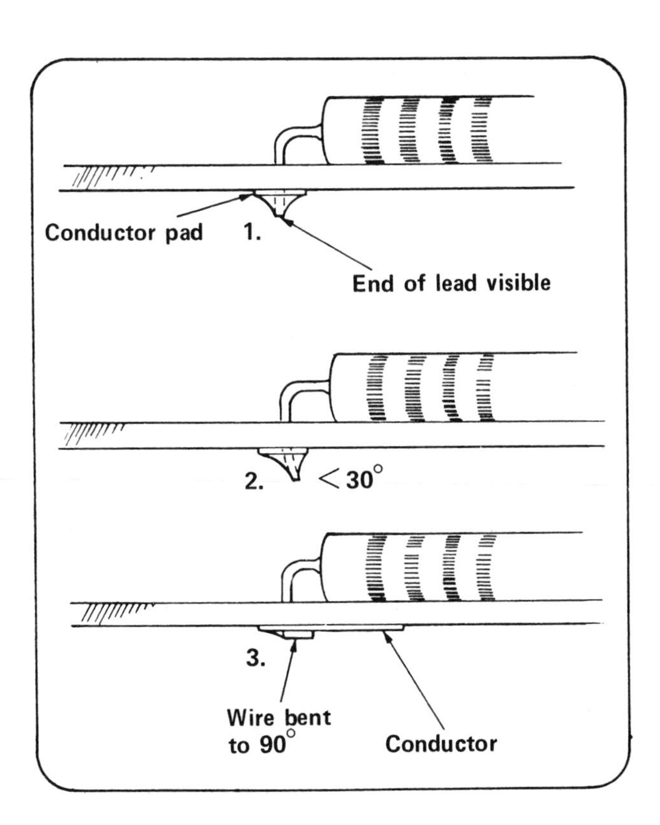

Good joints

Typical good joints are shown. The end of the lead should be visible through the solder and the surface of the solder should be concave.

1 The lead is not bent after insertion but remains at a right angle to the board; this provides fast insertion time, but less mechanical strength.

2 The lead is inserted and then bent to an angle of less than 30° from vertical. This holds the device in place.

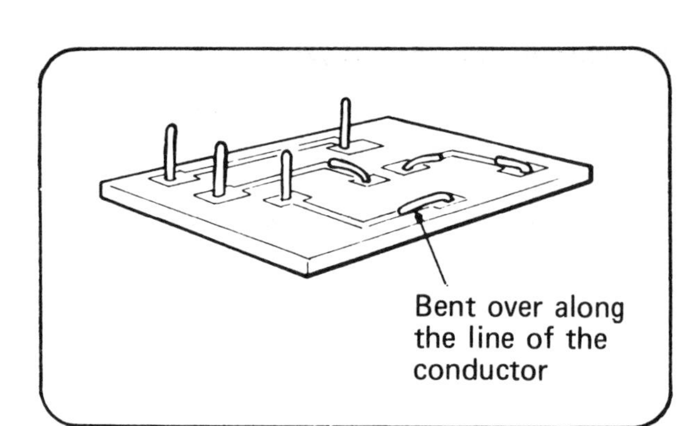

3 The lead is inserted bent parallel to the PCB. The lead is bent in the direction of the land.

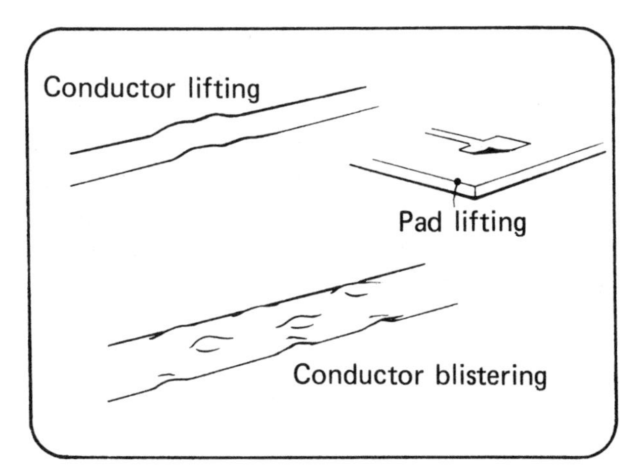

Bad joints

Bad solder joints can be caused if the iron is too hot. An overheated iron can cause the pads to lift, the conductor to lift and/or blister. The lifting of the pad is the result of the heat breaking the bond between the conductor and the board.

Inspection

Inspection is the final step. The inspector looks for problems such as

Pad lifting and conductor lifting or blistering.

Conductor breaks.

Excess flux.

Correctness of component values, layouts, and polarity.

Quality of the joints.

Wave-soldering

Wave-soldering is a mechanical method of soldering printed circuit boards, where all components are soldered at one time.

The components are assembled in position.

The leads are trimmed near the board and bent over along the line of the conductor to secure the components in position.

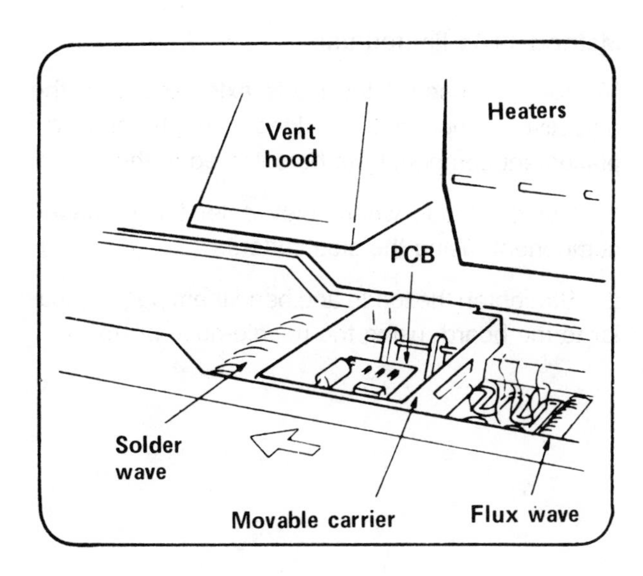

The board is loaded on a movable carrier.

The printed circuit board conductor strikes the top of a wave of flux. The board is then preheated before soldering. Soldering takes place when the conductor strikes the top of a wave of molten solder.

After soldering, the board is cleaned and then dried.

Removal and replacement of components

Components on a PCB may become defective, for instance, a fixed resistor may burn out. Often these defective components can be removed without removing the PCB from its chassis or rack mounting. The removal of components under these conditions is described in the following two methods. If the board is easily accessible, component removal is described in Method 3.

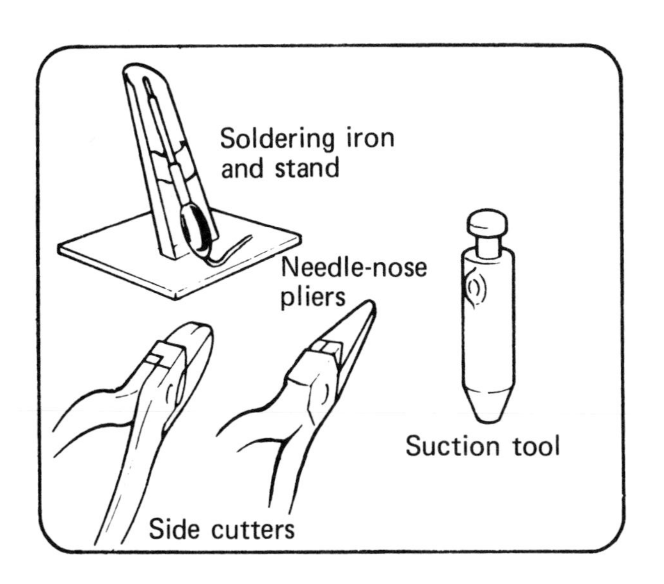

Tools

1 Soldering iron and stand.

2 Side cutters or diagonal cutters.

3 Needle-nose pliers.

4 Suction tool for removing solder or desoldering wick.

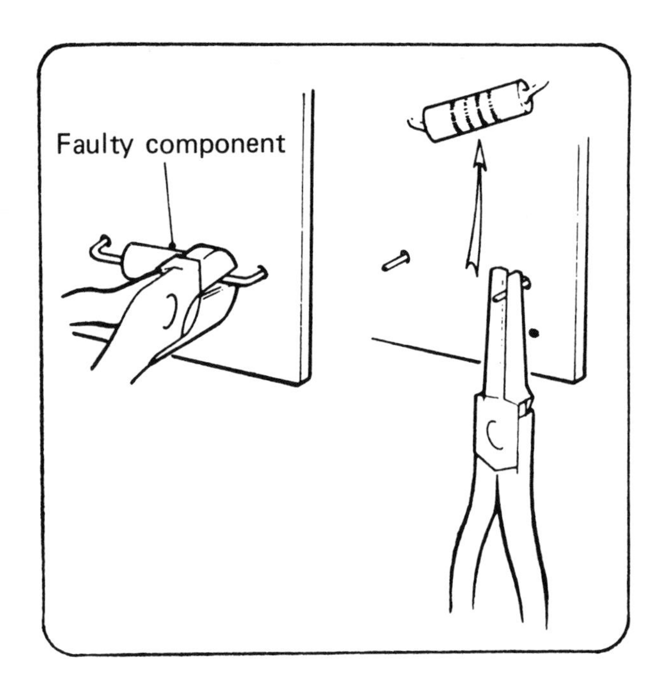

Method 1—In limited use

Method 1 is used if the leads extending from the defective component are long enough for a replacement component to be soldered to them.

1 Cut the leads where they enter the damaged component, using the side cutters.

2 Straighten the leads and bend them perpendicular to the board, using the needle-nose pliers.

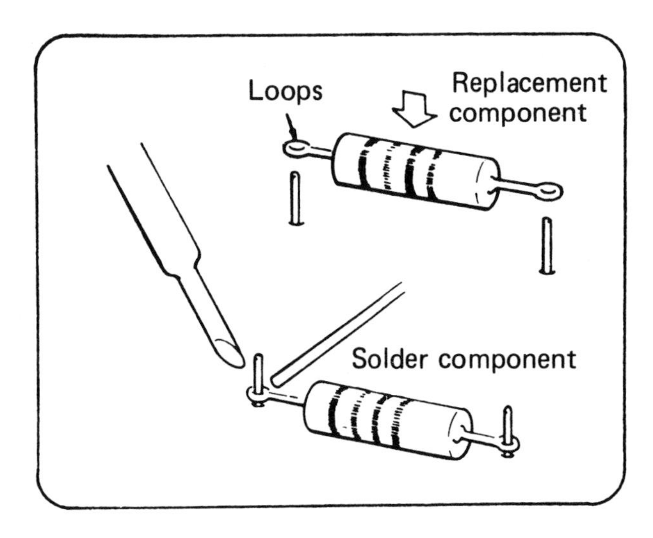

3 Cut the leads of the replacement component and bend a small loop in each end.

4 Place the loops over the old component leads.

5 Solder the replacement component in position.

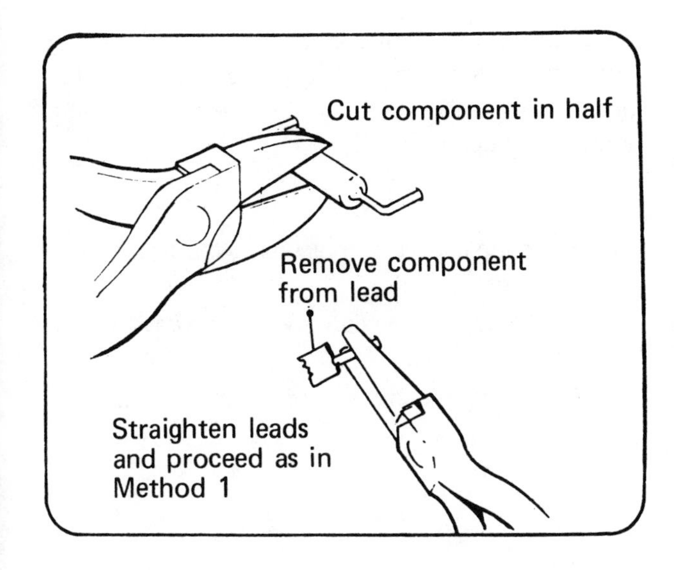

Method 2

Method 2 is used if the leads extending from the defective component are not long enough to use Method 1.

1 Cut the defective component in half, using the side cutters.

2 Remove the broken halves from the leads and proceed as in Method 1. This is used to obtain as much lead length as possible to solder the new component to the old leads.

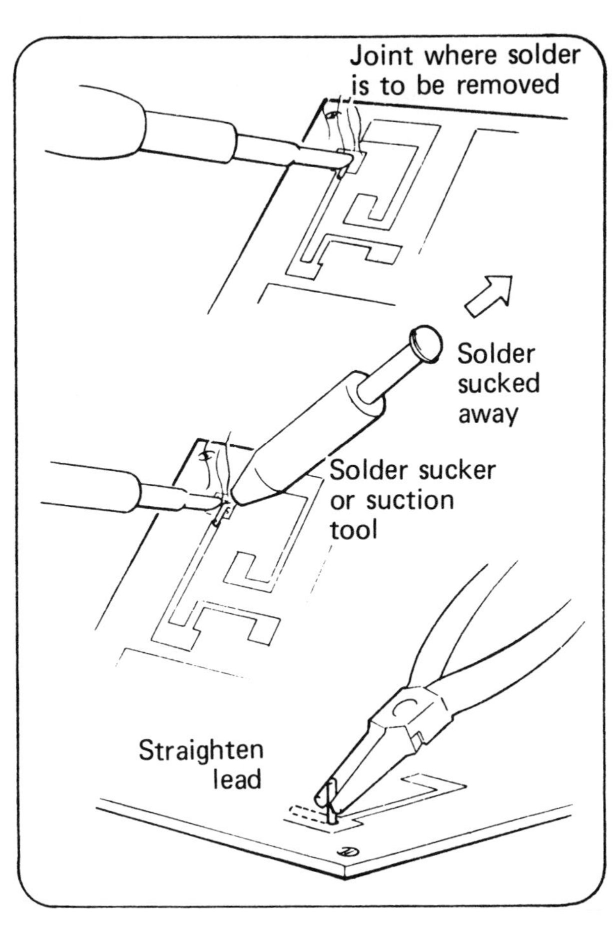

Method 3

Method 3 is used when the board is easily accessible.

1 Heat the connection on the conductor side of the board using the soldering iron.

2 Use the solder sucker as follows: Heat the solder until it is molten; place the nozzle of the suction tool against the solder; press the button. This allows the plunger to spring out and the solder is sucked away.

3 If the lead has been bent over, straighten it with the pliers.

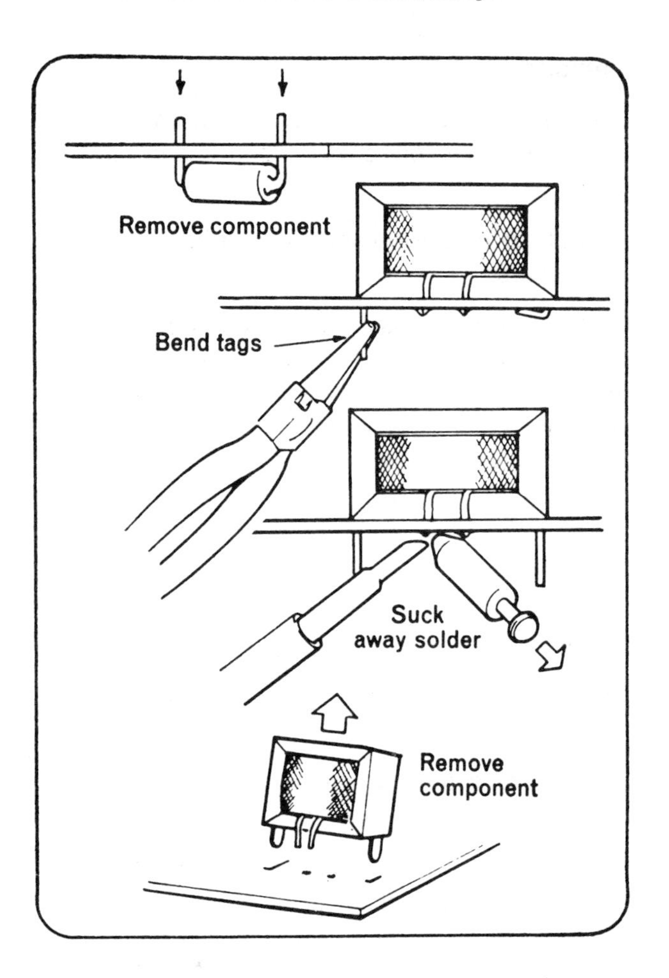

4 Unsolder the other connection (or connections) and remove the component. For components that are held in position by bent over metal tags (transformers, inductors, etc.).

(a) Bend the tags perpendicular to the board using the pliers.

(b) Heat each connection and suck away the solder.

(c) Remove the component from the board.

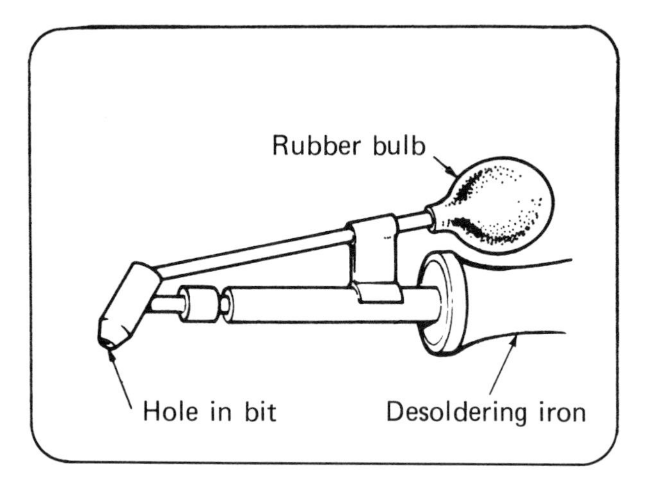

Desoldering iron

A desoldering iron can be used for melting and sucking away solder as follows.

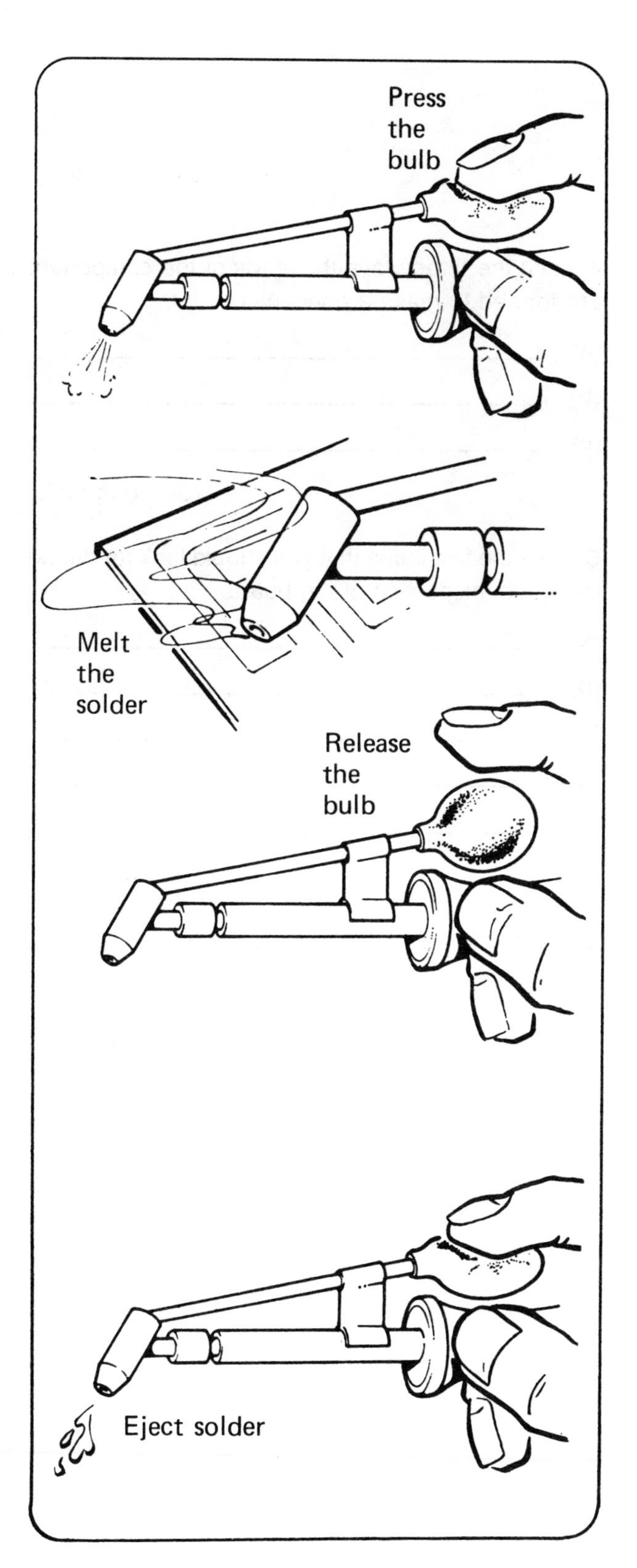

1 Press the rubber bulb to suck out most of the air.

2 Melt the solder with the tip of the iron.

3 Release the bulb. The molten solder will be sucked away through the hole in the tip.

4 Remove the solder from the desoldering iron by pressing the bulb immediately after use.

Test items

1 Which one of the following drawings would tell you the order components should be mounted on the PCB?

(a) Parts list.

(b) Component layout.

(c) Sequence chart.

(d) Schematic.

(See page 25.)

2 Which of the following items should you *not* do when working with PCBs?

(a) Handle the board with sweaty hands.

(b) Stack the boards on top of each other.

(c) Flex the boards.

(d) All of the above.

(See page 26.)

3 Low wattage resistors are normally mounted _______ with the board.

(See page 29.)

4 List the three ways the leads of the component are formed to make a good joint.

(a) ________________________________

(b) ________________________________

(c) ________________________________

(See page 31.)

5 List the five items that you should look for during the visual inspection of the board.

(a) ________________________________

(b) ________________________________

(c) ________________________________

(d) ________________________________

(e) ________________________________

(See page 32.)

6 Describe why a desoldering iron would be used.

____

(See page 36.)

Module four

Double-sided printed circuit boards

Introduction

This module covers techniques associated with the manufacture of double-sided printed circuit boards. The process of making interconnections by using plated-through holes is described. The photo resistive process of etching the board is completely outlined. Safety precautions are reviewed.

Key technical words

- **Datum** A reference point or hole.
- **Plated-through-hole** A hole that has been coated with a conductive material.

Performance objectives

Upon completion of this module, you should be able to:

- Explain why a plated-through hole is used.
- Describe why a datum hole is used.
- List the steps in the plating process.
- List the steps in preparing a PCB for the photo resist process.

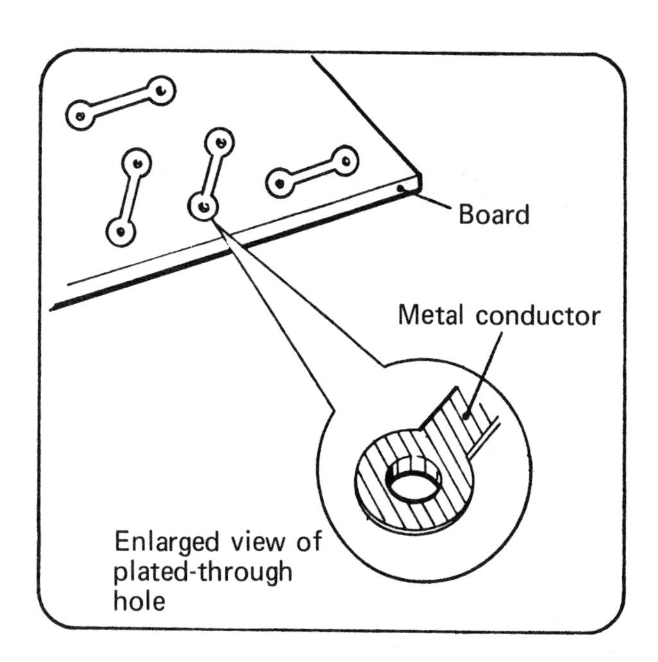

Double-sided boards

Plated-through holes

A double-sided printed circuit board has copper foil on both sides of the insulating board. It is often necessary to connect one side of the board to the other side. A plated-through hole has a continuous layer of metal lining the walls of the holes.

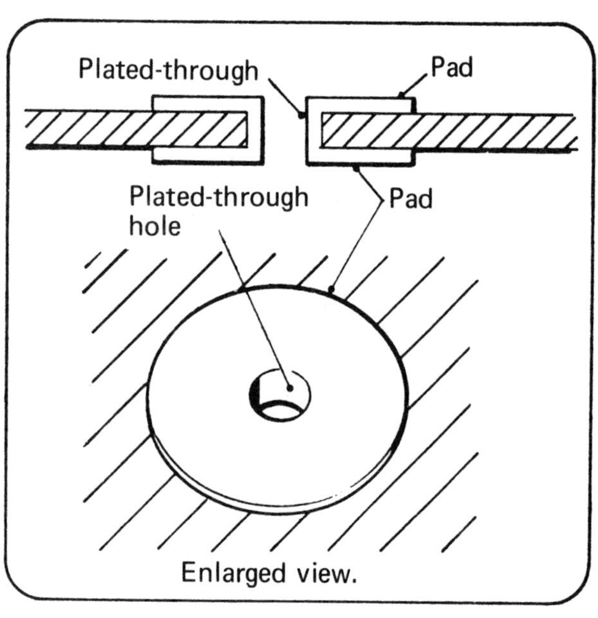

The plated-through holes provide electrical connections between the printed circuits on each side of the board.

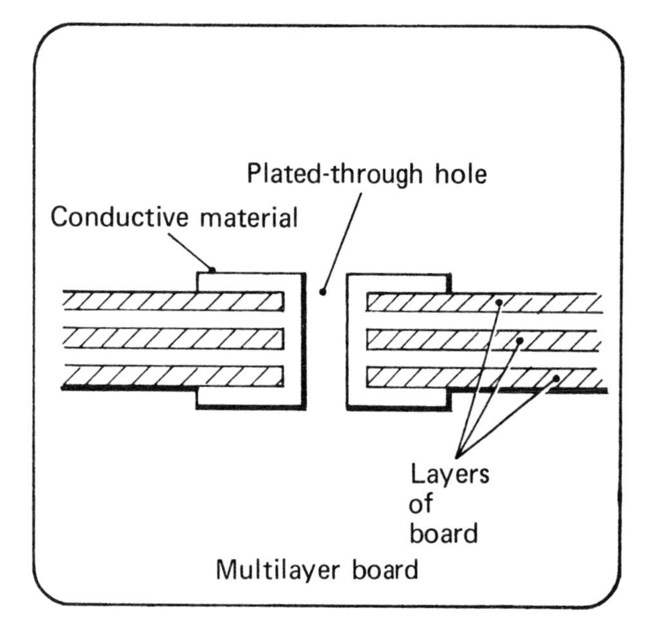

A plated-through hole can also provide electrical connections between the layers of multilayer boards.

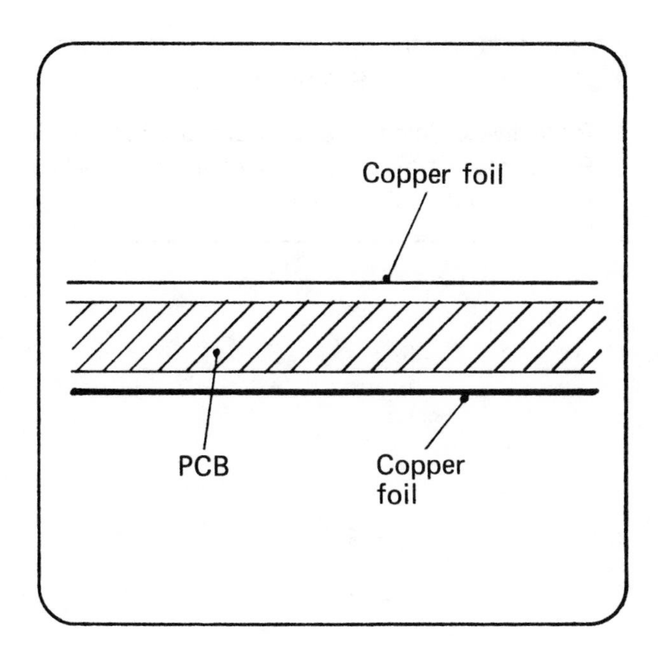

Method of production

Raw material

Before processing, the PCB consists of a sheet of epoxy/glass fiber board or other suitable material clad on both sides with a thin layer of copper.

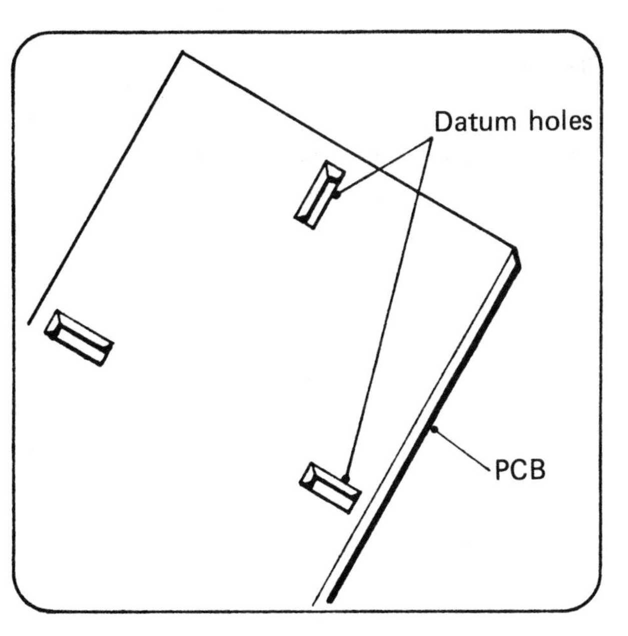

Setting a datum

The PCB is cut slightly larger than the required size of the finished board. Holes are cut into the board for use as *datum* points or reference points. These data points are used throughout the process to ensure that all pads, conductors, holes, and edge connectors are correctly positioned.

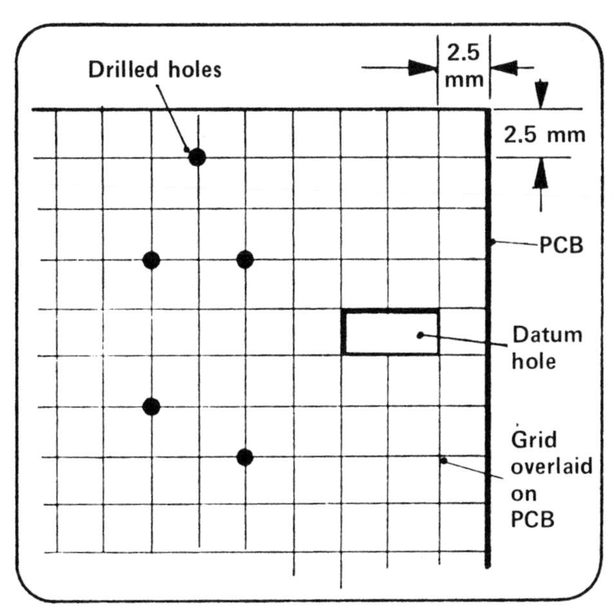

Grid system

Holes are drilled through the board where specified. These holes conform to a 0.1 in. (2.5 mm) grid system. On subsequent process operations, this grid system is used, insuring that the pads are accurately positioned over the holes. The holes are drilled on the intersection of two lines of the grid. Component leads are subsequently formed to the grid sizes and will fit accurately to the specified pad holes.

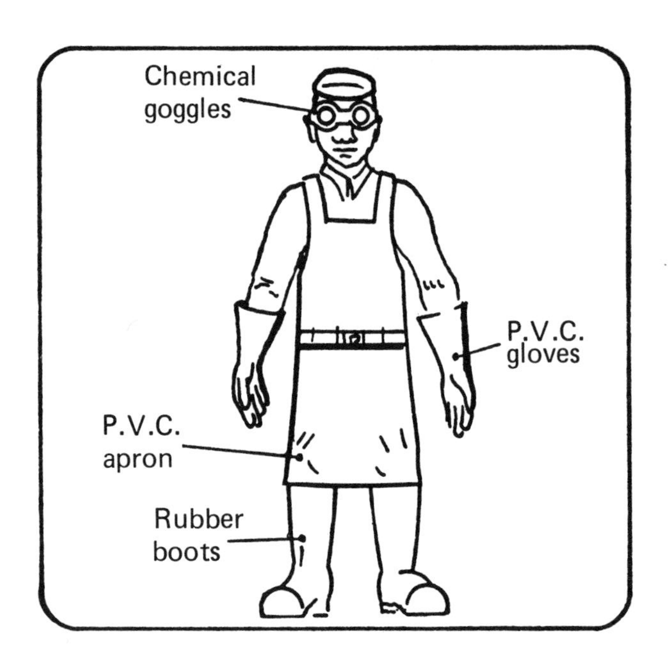

Safety

Protective clothing must be worn at all times. PCBs must be handled carefully to avoid splashing harmful chemicals.

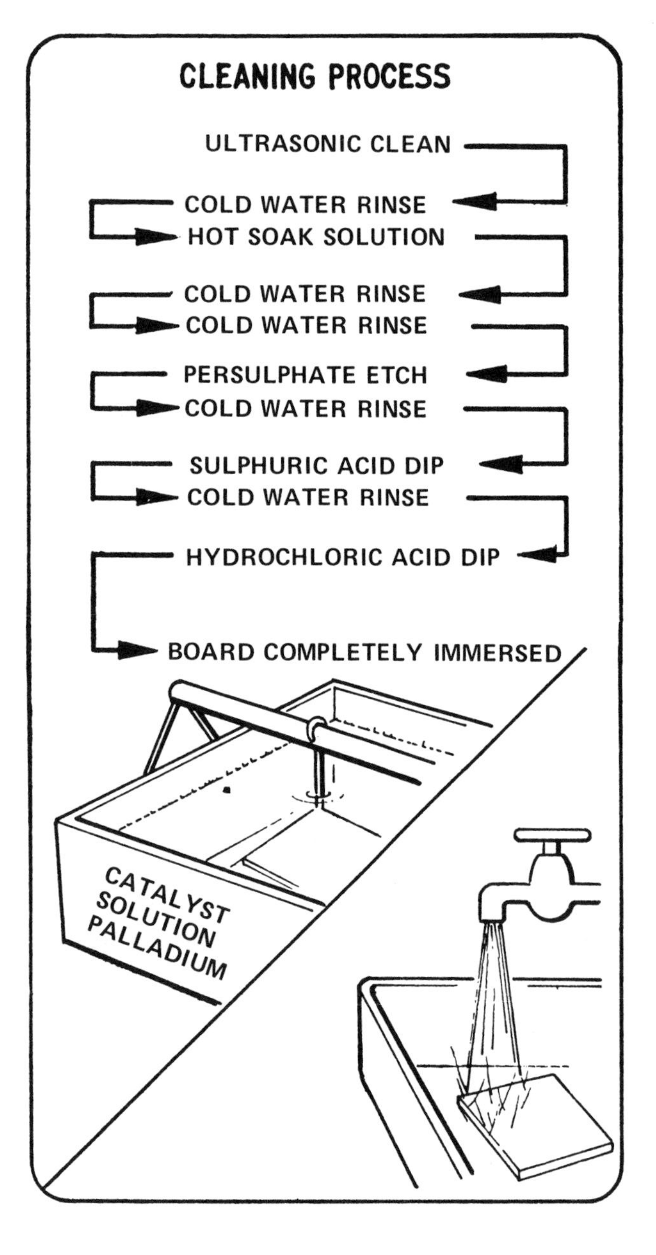

Cleaning

The cleaning process ensures that no impurities remain on the board. Once cleaned, the board is ready for metallizing or plating.

Metallizing—Stage one

The board is immersed in a catalyst solution in which minute particles of palladium are suspended. These particles become attached to all the exposed surfaces of the board, thus forming a very thin layer of palladium. The board is then thoroughly rinsed twice in cold running water.

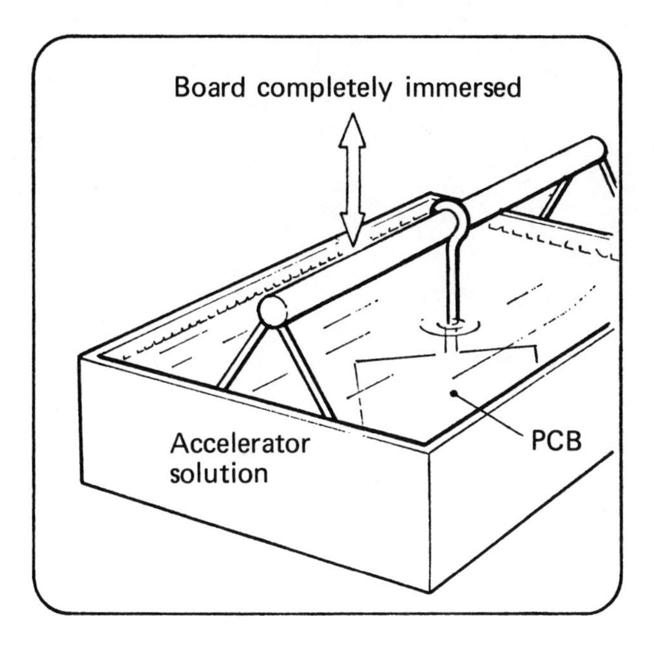

Accelerator

The board is immersed in a special solution called an accelerator, for a specified period. Certain chemicals are removed, leaving a thin palladium layer ready for the next stage of the process.

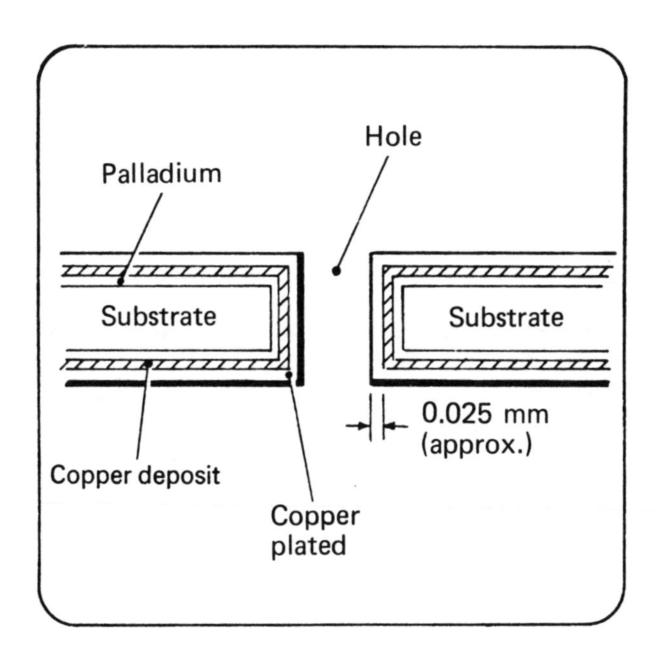

Metallizing—Stage two

The board is immersed in an "Electroless" copper plating solution.

Copper is deposited over the palladium in order to produce the firm base necessary for electroplating. This plating is extremely delicate at this stage and varies in thickness between 0.00002 and 0.00005 in. (0.5 and 1.25 μm).

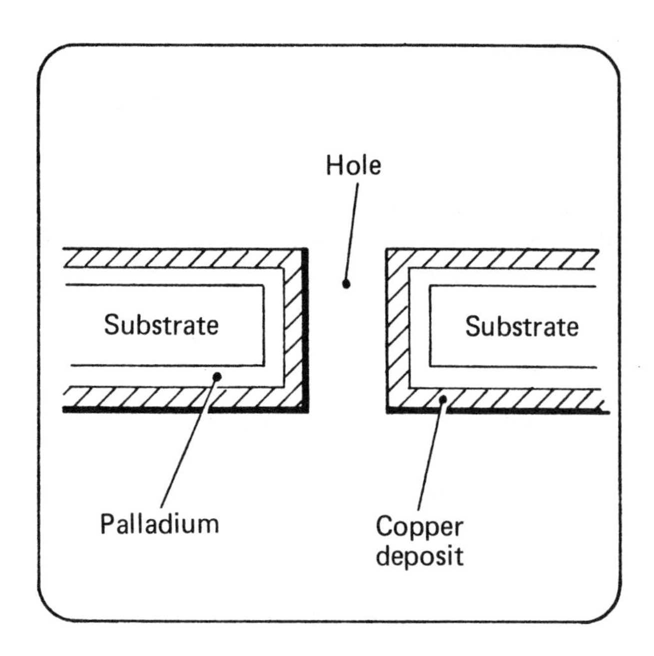

Electroplating

The electroplating process is carried out, building up the copper coating to a thickness of about 0.001 in. (25 μm). These steps and dimensions may vary depending upon special circuit requirements. The above steps show how a plated-through hole can make connections to both sides of the board.

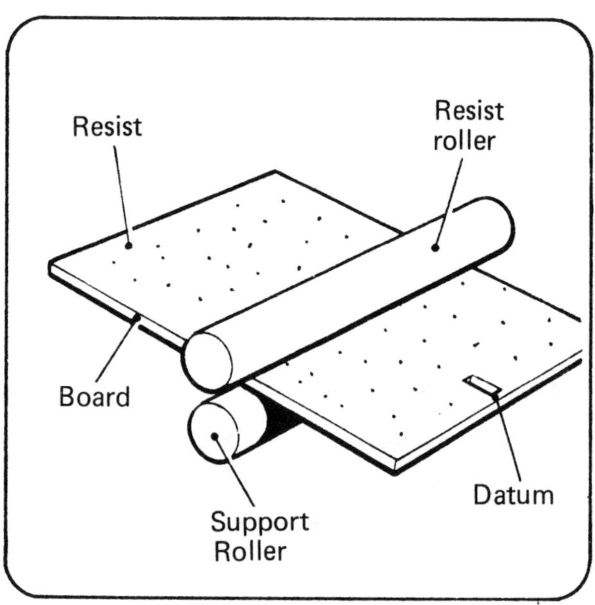

Photo resist

The board is covered on both sides with a light sensitive layer called a resist. Various methods are used for putting on the resist, for example, by roller, which deposits a soft, even layer onto one side of the board.

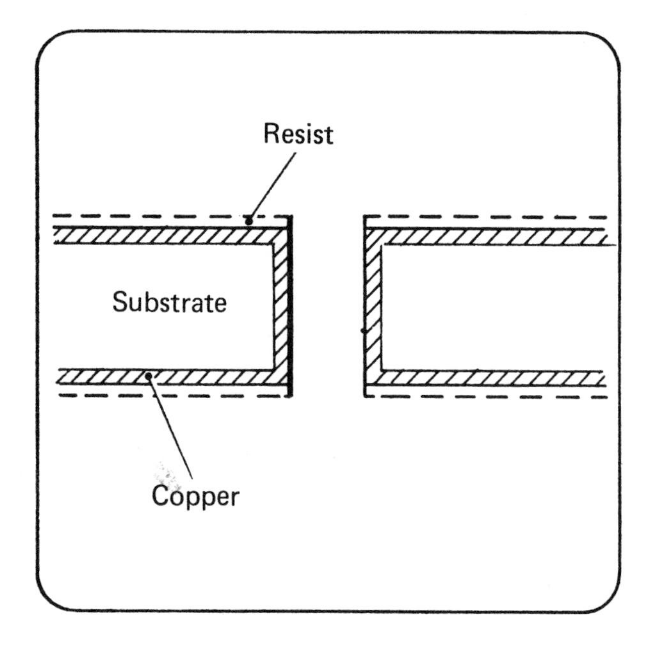

This resist is allowed to dry and the operation is then repeated for the opposite side. Resist will become insoluble to solvents ("hardens") when exposed to ultraviolet radiation. The resist is light sensitive (UV radiation) at this time. The board should be handled in subdued lighting or darkroom conditions.

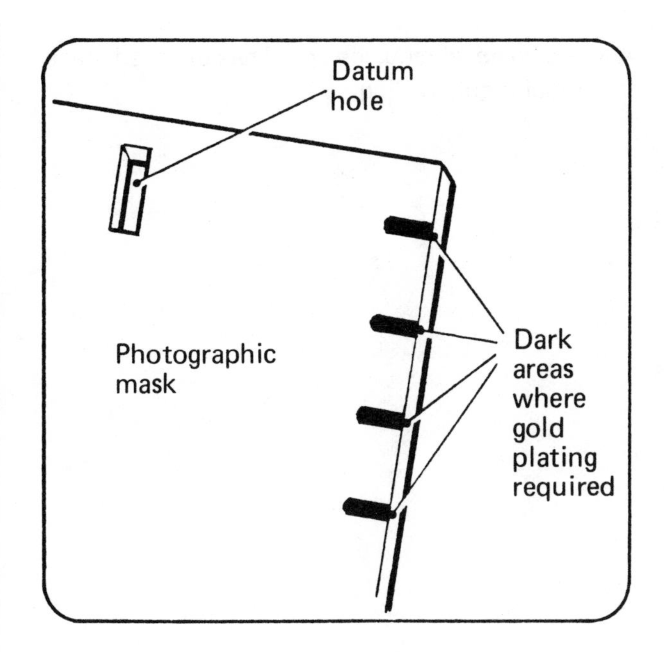

Gold contacts

A photographic mask is prepared from a scale drawing of the gold contacts. The photographic mask has dark areas in positions corresponding to the gold contacts required on the board.

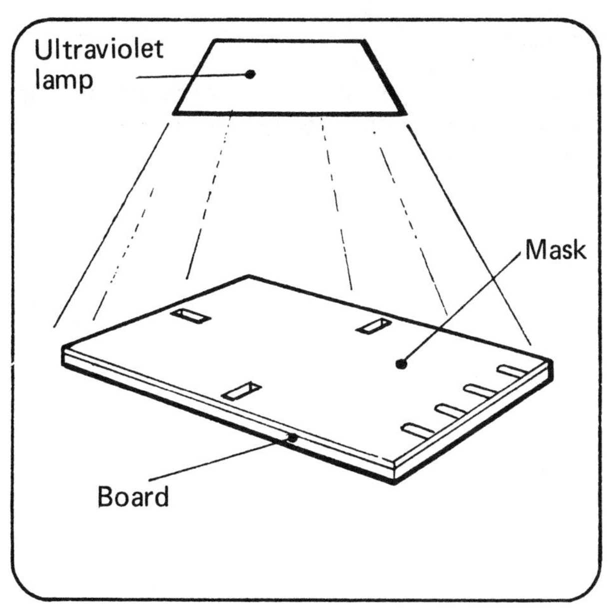

Exposure

The photographic mask is placed in contact with the resist on the board and is then exposed to ultraviolet radiation. This hardens all resist on one side of the board, except that shaded by the mask. Both sides of the board are treated this way.

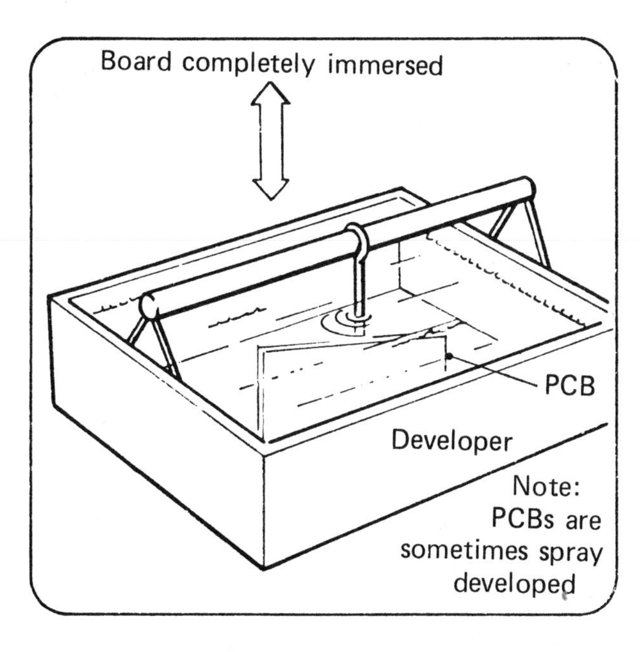

Developing

The board is now immersed in a liquid that dissolves unhardened resist. The liquid is called a developer.

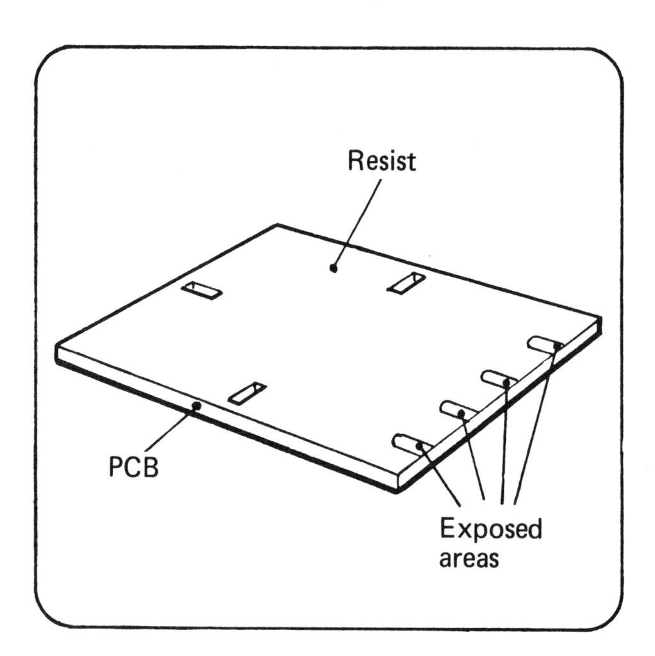

After being developed, the board has exposed areas that require gold plating.

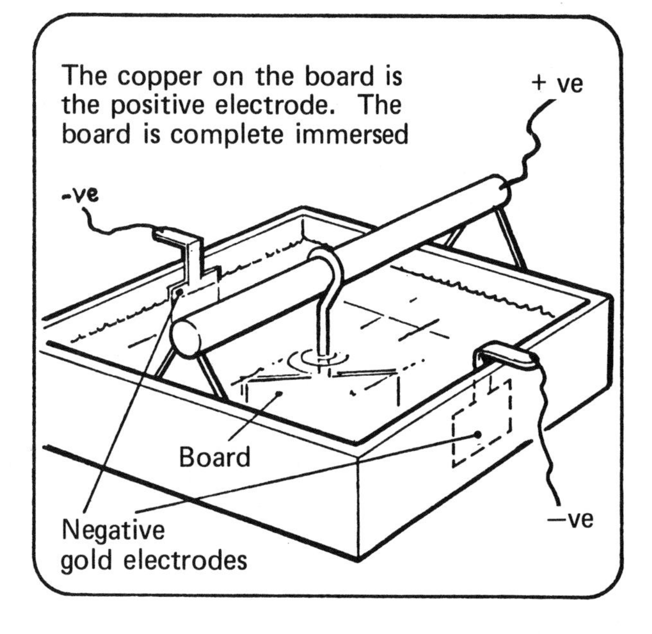

Gold plating

Gold is electroplated onto the exposed areas of the board. Gold has higher conductivity, less oxidation, and less wear than copper.

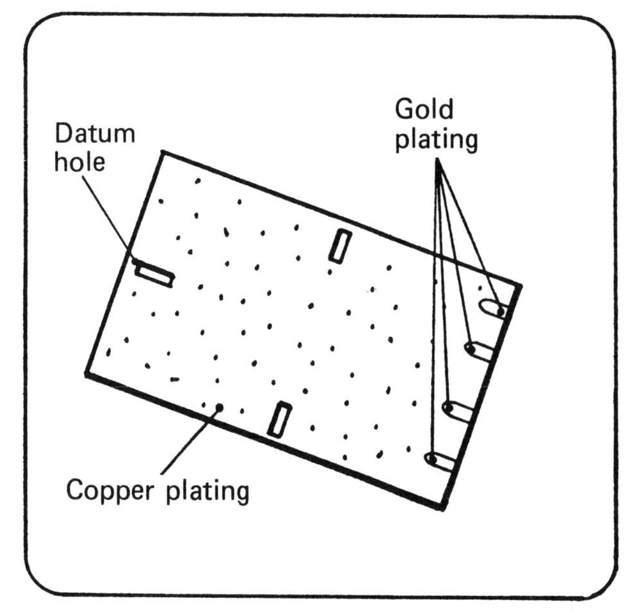

All resist is chemically stripped from both sides of the board. The board is left with areas of gold on the copper plating.

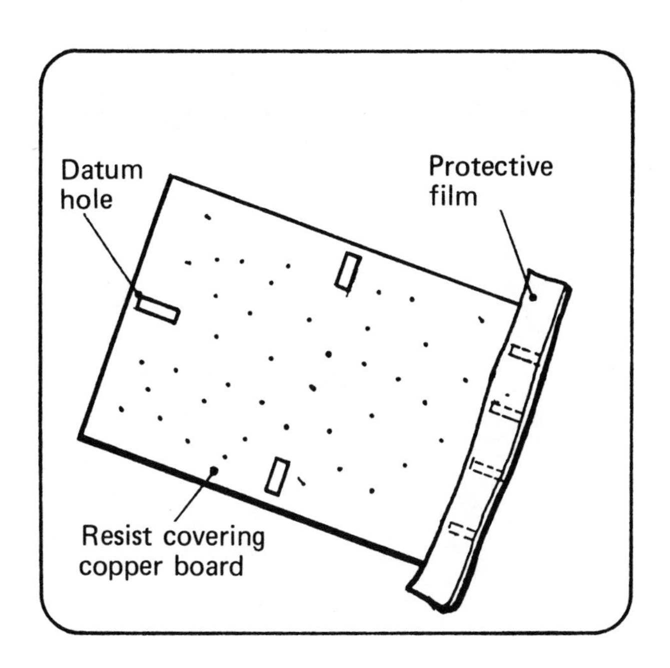

The gold areas are then covered with protective film and fresh resist is applied to both sides of the board.

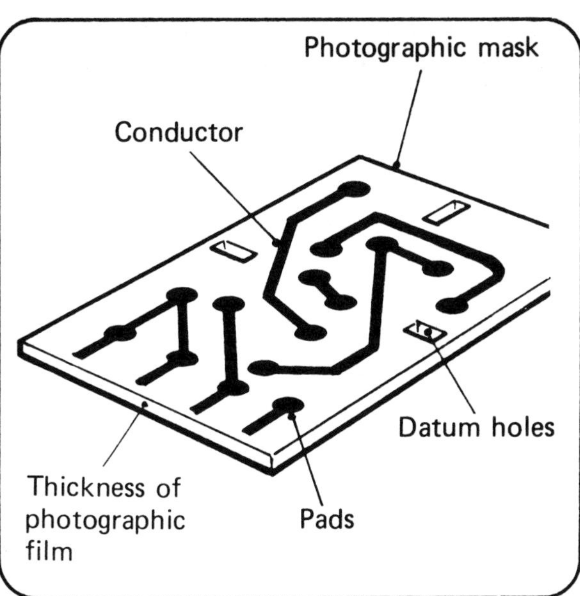

Conductor and pads

A photographic mask is prepared from a drawing of the conductor pattern. The mask has dark areas that correspond with the conductor and pad positions required on the board.

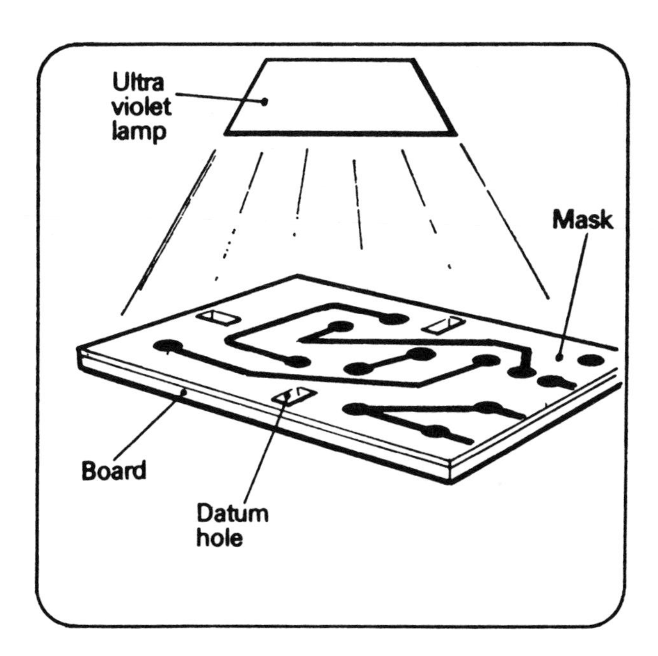

Exposure

The photographic mask is placed in contact with the resist of the board. The board is then exposed to an ultraviolet lamp. This hardens resist not covered by the dark areas of the mask. Both sides of the board are treated in this way.

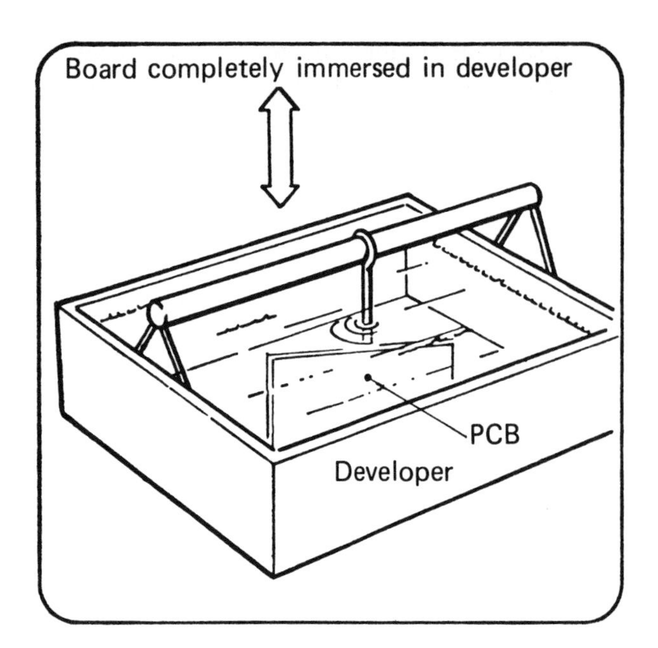

Developing

The developer is used to dissolve all unhardened resist.

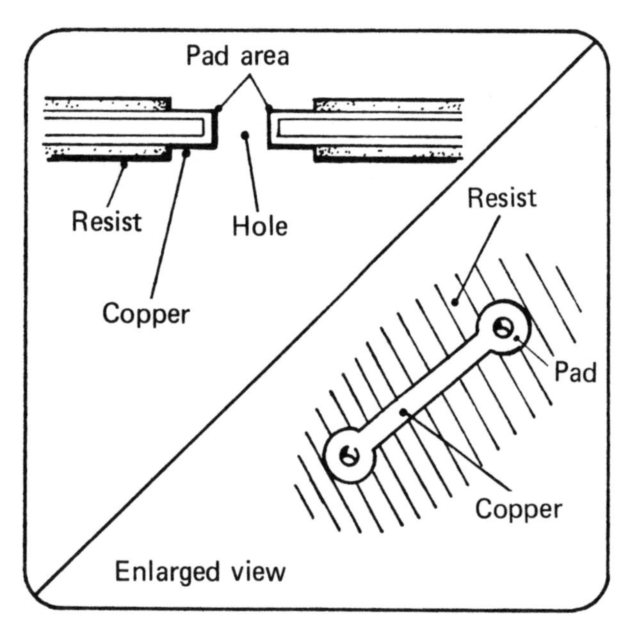

The board now has bare areas of copper surrounded by resist on both sides of the board.

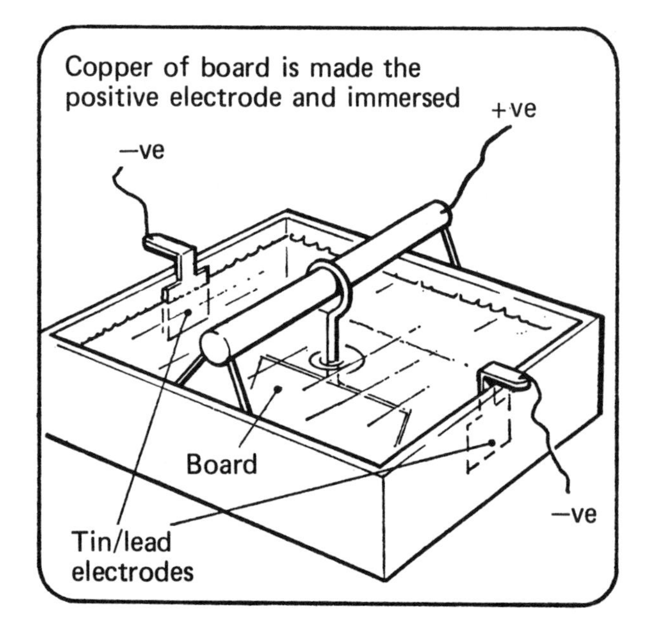

Tin/lead plating

A tin/lead alloy is electroplated onto the bare copper of the board, including the inside of the hole. This increases solderability and reduces oxidation of the copper.

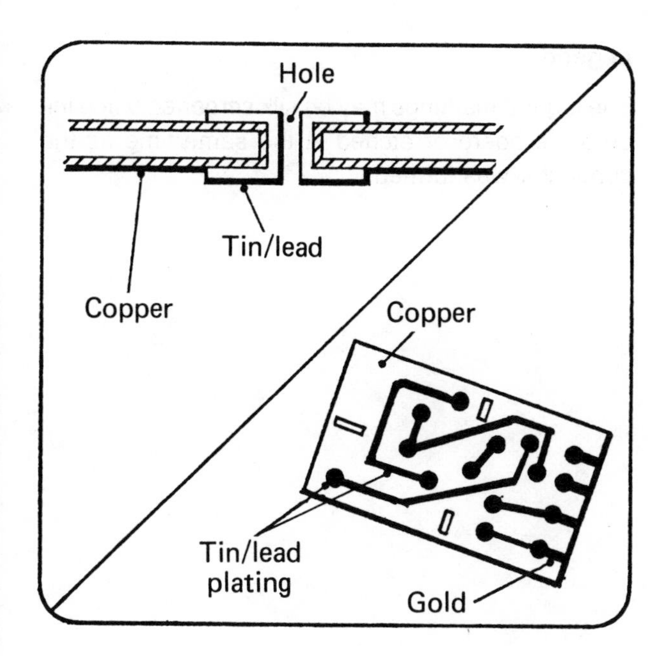

All resist is chemically stripped from both sides of the board. This leaves areas of copper coated with tin/lead or gold plating.

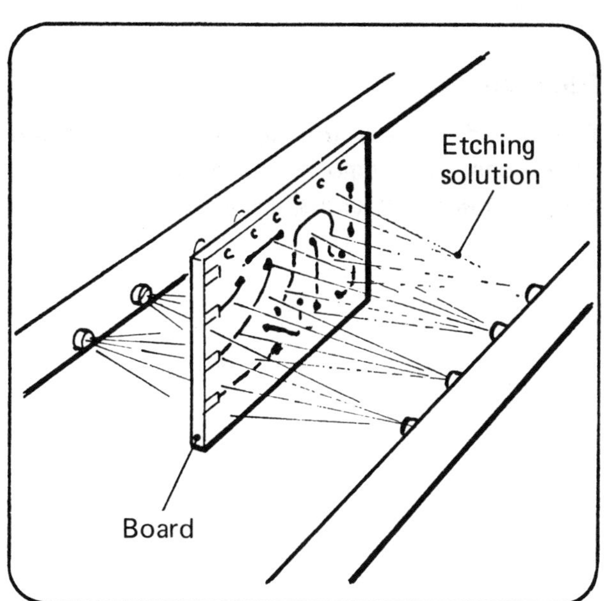

Etching

An etchant is sprayed onto both sides of the board to remove unwanted copper. A chromic acid/sulphuric acid mixture is commonly used. It removes copper but leaves the gold and tin/lead intact.

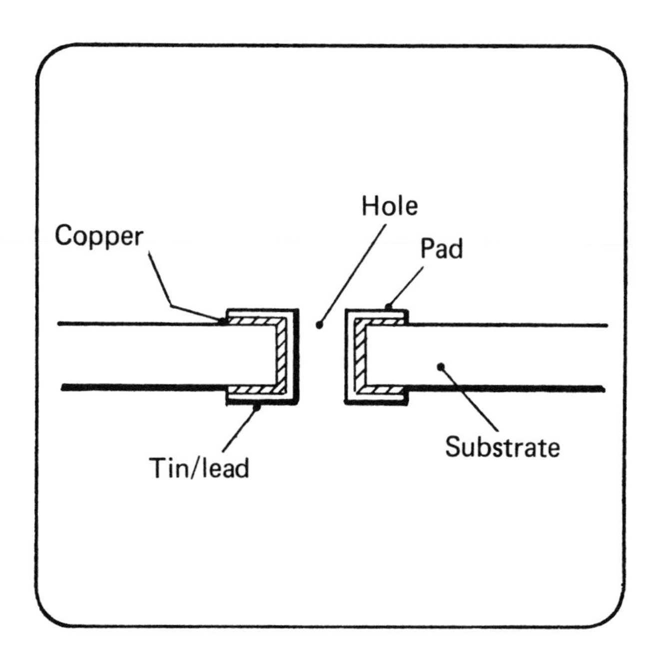

After thorough rinsing in cold water, the conductors, pads, and gold contacts are clearly defined on the board.

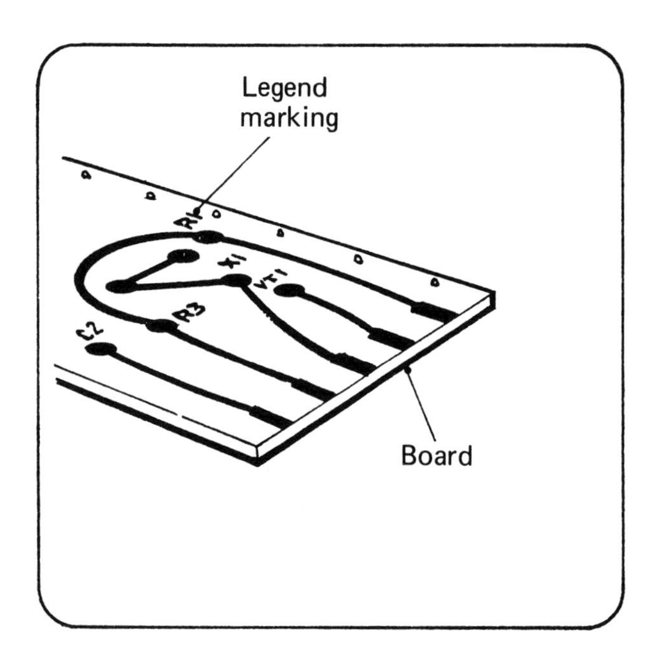

Legend

The legend markings may be silk screened using ink onto the board or etched at the same time as the conductor was formed.

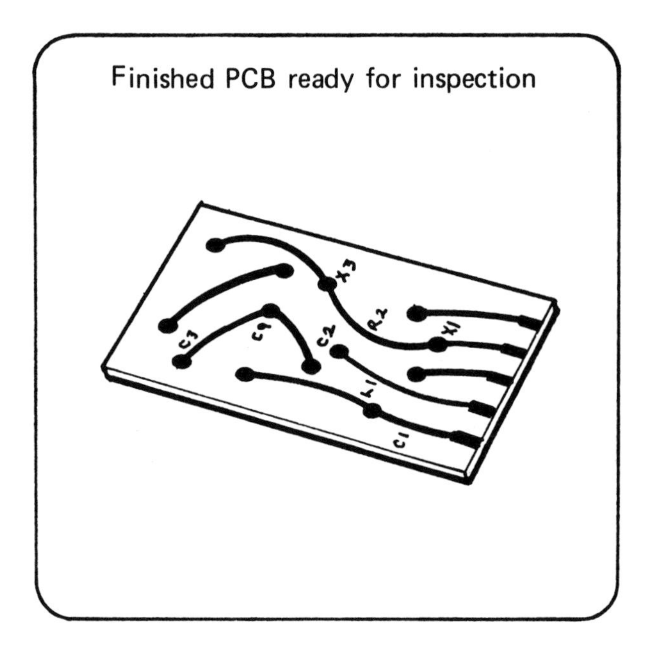

Trimming

The board is now trimmed to the specified dimensions. This removes the datum holes. The board is now ready for inspection.

Test items

1 A ______ - ______ ______ is used to connect one side of the board to the other side of the board.

(See page 40.)

2 A ______ is used to provide a reference point for all the drawings, photographic masks, drilling alignments, etc.

(See page 41.)

3 Describe what protective clothing is required in handling the harmful chemicals.

(a) ______________________

(b) ______________________

(c) ______________________

(See page 42.)

4 What type of light source is used to expose the photoresists material?

____

(See page 45.)

5 State why gold is plated onto the edge connectors.

____

(See page 46.)

6 Name a commonly used etching solution.

____

(See page 8.)

Module five

Multilayer printed circuit boards

Introduction

This module examines the uses of multilayer printed circuit boards and the steps required to manufacture a multilayer board. Handling precautions required for multilayer boards are the same as those required for single and double-sided boards. These precautions are reviewed.

Final testing and inspection techniques for multilayer printed circuit boards are discussed. The complexity of multilayer boards requires great care in all stages of manufacturing and assembly.

Key technical words

- **Multilayer printed circuit board** Alternate layers of copper foil and insulating material combined to make a very compact circuit board.
- **Interconnection** An electrical connection from one layer to other layers in a multilayer printed circuit board.

Performance objectives

Upon completion of this module, you should be able to:

- State the advantages of a multilayer printed circuit board.
- Identify common faults associated with multilayer printed circuit boards.
- Explain the purpose of the glass/epoxy resin used in the manufacturing of a multilayer printed circuit board.

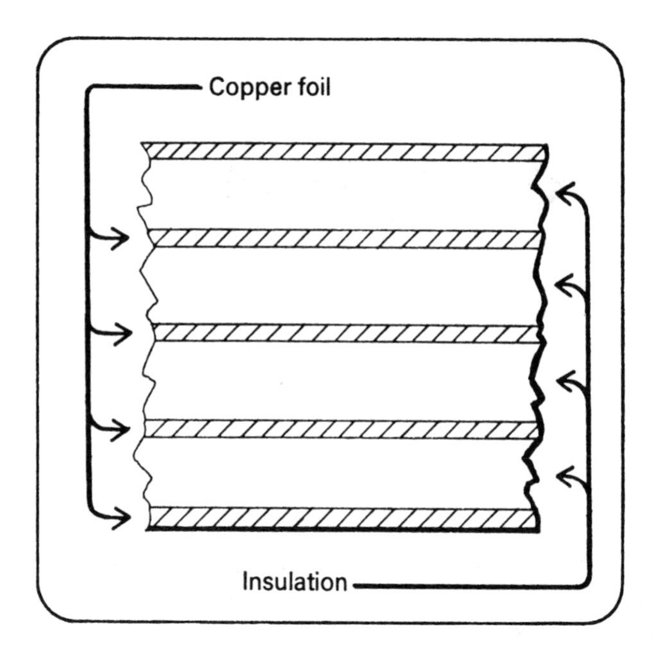

Multilayer printed circuit boards

Multilayer printed circuit boards consist of alternating layers of copper foil and insulating material. There may be many layers in one board and they are very closely packed. The total thickness of the board is usually within the range 1/32 to 3/32 in. (0.700 to 2.400 mm).

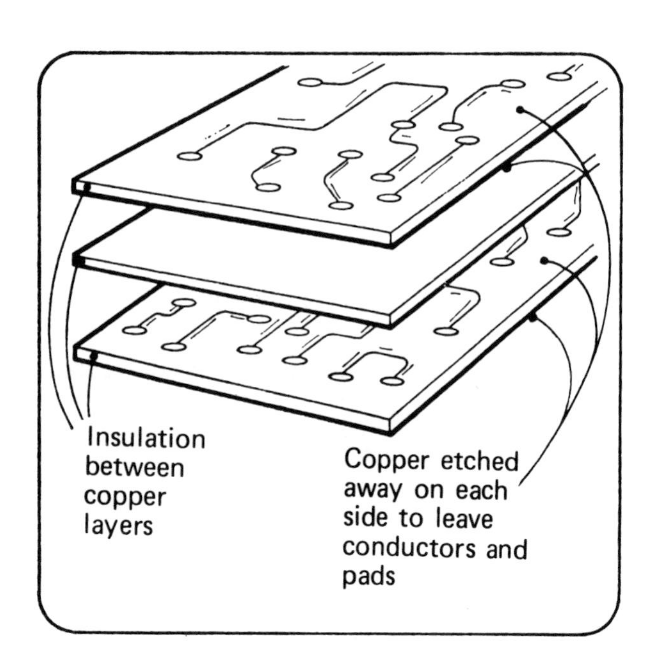

In the same way that parts of the copper foil were etched away to leave conductors and pads on an ordinary printed board, the manufacture of multilayer boards is accomplished at each *buried* layer of copper to leave printed circuit wiring sandwiched between the insulating layers.

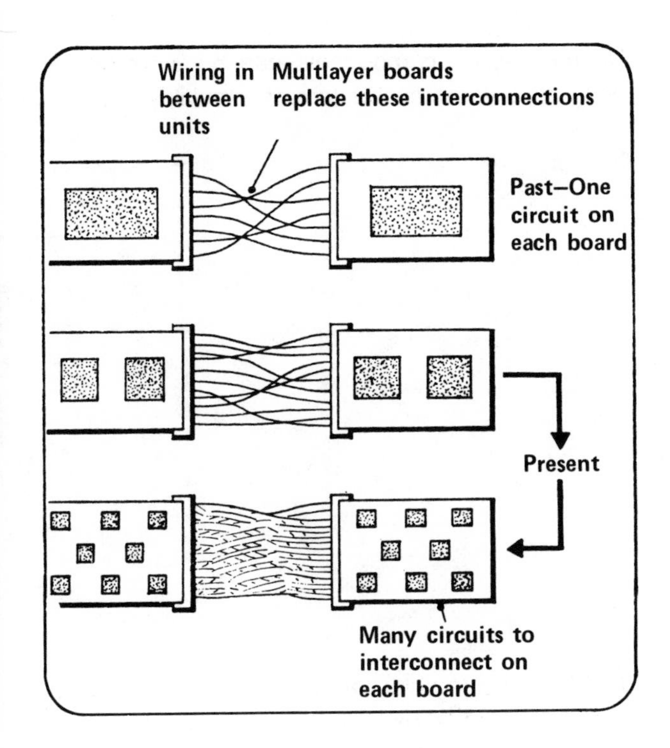

Uses of multilayer printed circuit boards

Multilayer printed circuit boards have many applications. Among the most significant are

1 The replacement of complex interunit wiring to make the equipment lighter, smaller, and cheaper.

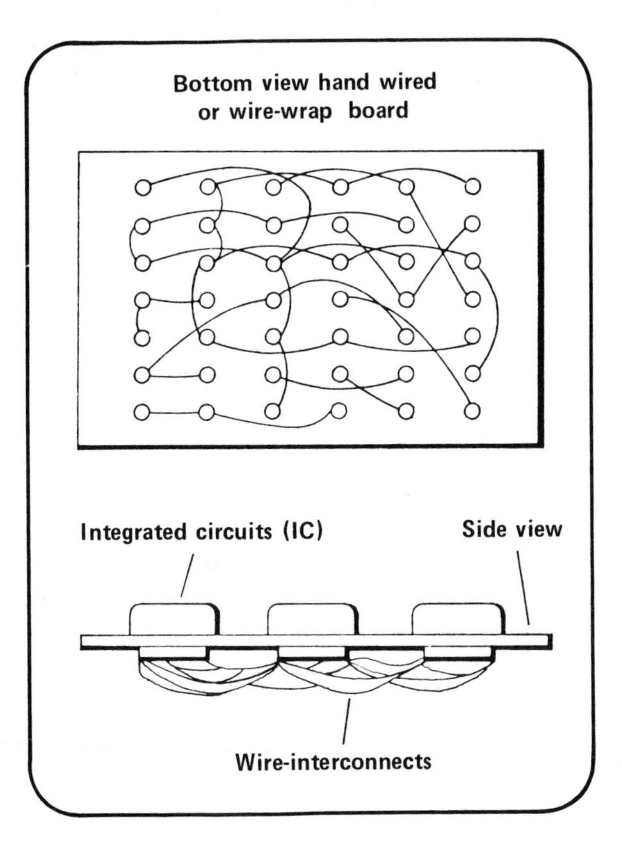

2 The accommodation of closely packed components in the smallest possible space. This is important because of the trend towards smaller components and, in particular, integrated circuits. (The illustration shows conventional wiring between three microelectronic devices.)

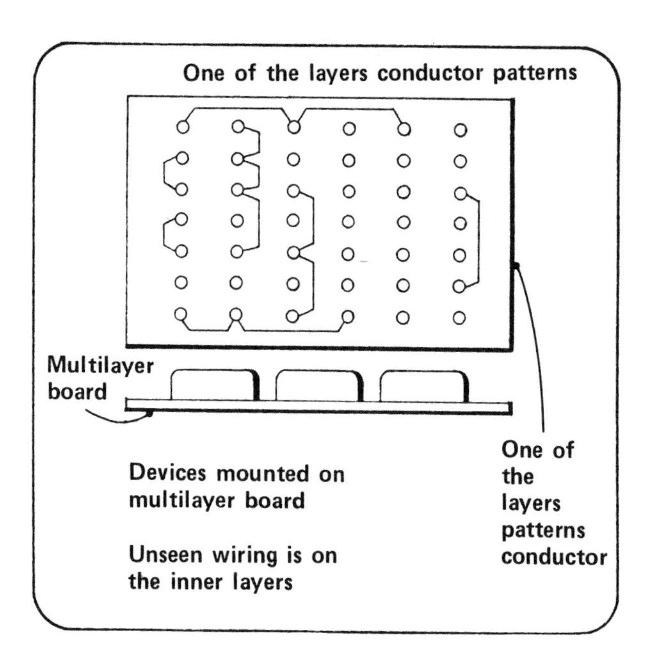

Closely packed components can be achieved by using the layered conductors of a multilayer board.

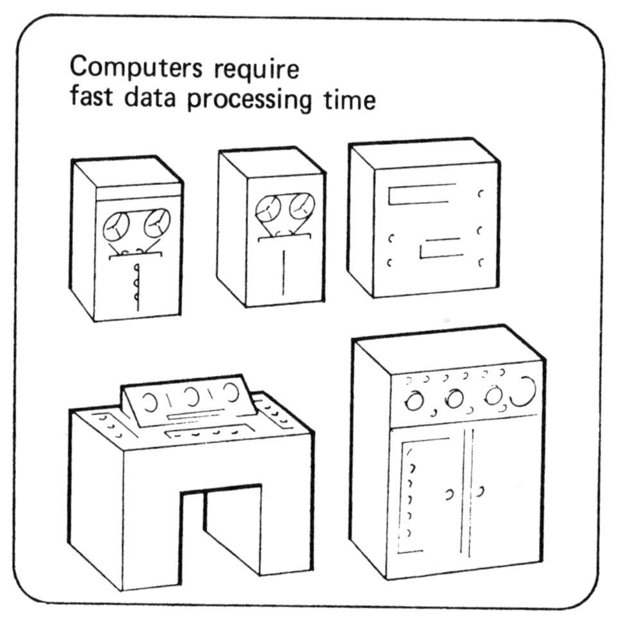

3 The considerable decrease in the time that an electrical pulse takes to travel from one component to another is a result of the *shorter* wiring that is possible with these boards. This is a very important factor in many types of equipment such as computers, where the speed of information processing must be as fast as possible.

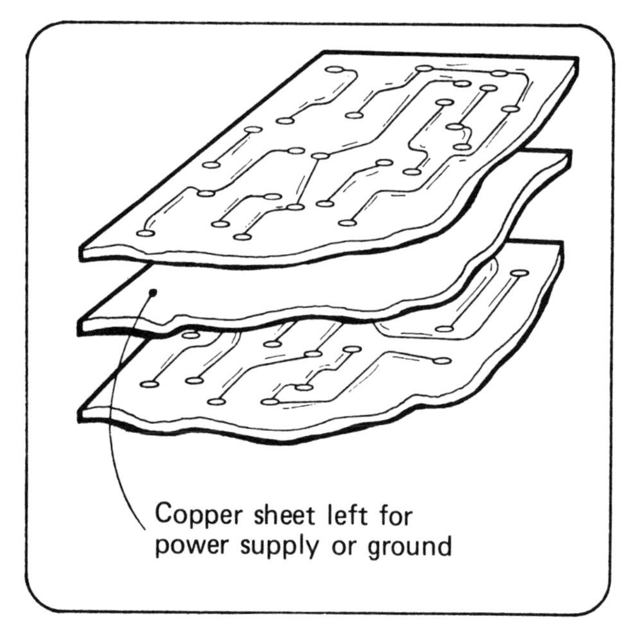

4 Some of the layers may be left as almost complete sheets of copper foil and these can be used for grounding or to supply power in such a way that a connection may be made to them anywhere over the area of the board.

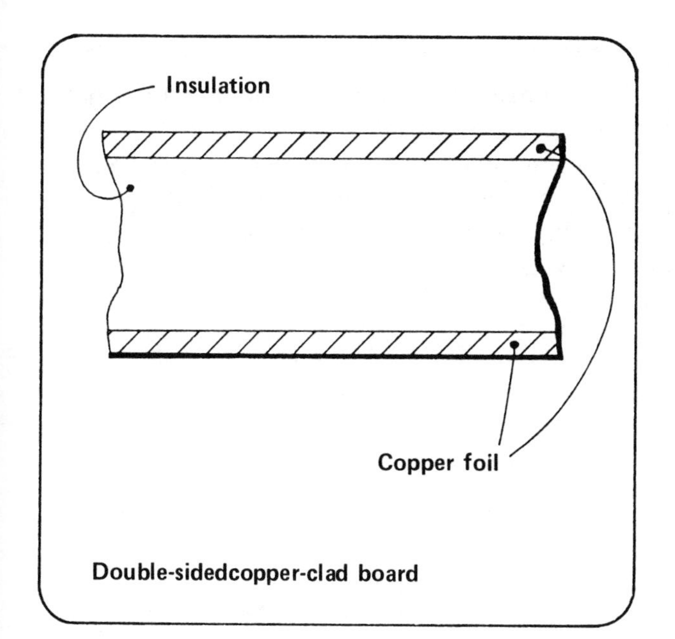

Double-sidedcopper-clad board

Method of manufacture

A multilayer board is made by joining together a number of thin glass/epoxy resin boards or boards of similar material. Each one is covered on both sides with thin copper foil and is known as a cured double copper-clad board. The following process relates to glass/epoxy resin although other materials can be used.

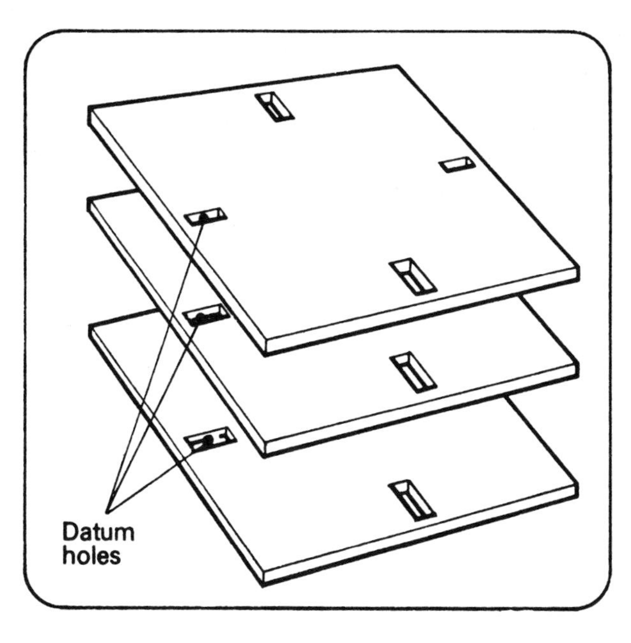

Stage 1

Accurate datum holes are made in each double copper-clad board. These insure that the boards are correctly positioned when they are joined together.

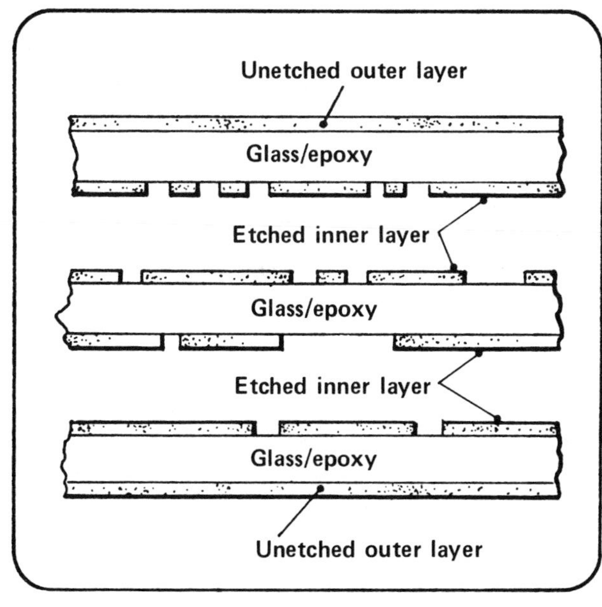

Stage 2

Each copper foil that will eventually form an inner layer is separately printed and etched with its own wiring pattern. (The method is the same as that for ordinary printed circuit boards.) The copper that forms the two outer layers is not printed or etched at this stage.

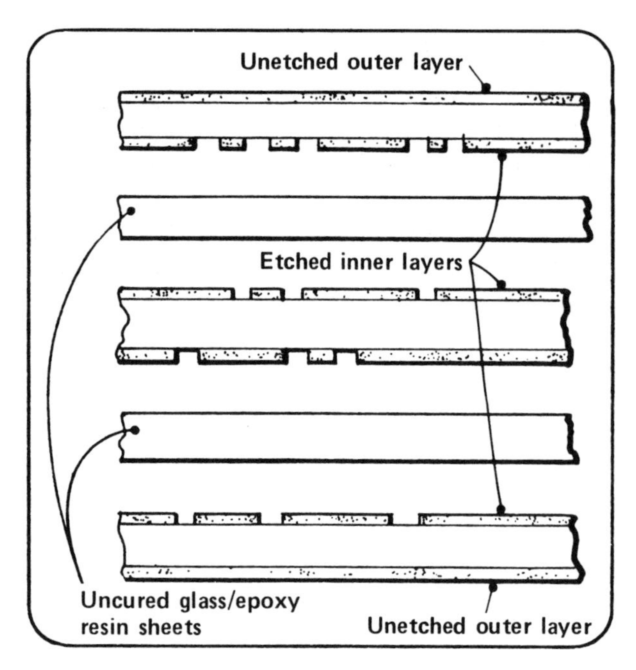

Stage 3

More sheets of glass/epoxy resin then have holes cut into them, similar to the datum holes. These sheets are different—they are not copper-clad and they have not been cured. They will become a bonding agent and insulation between the two copper layers.

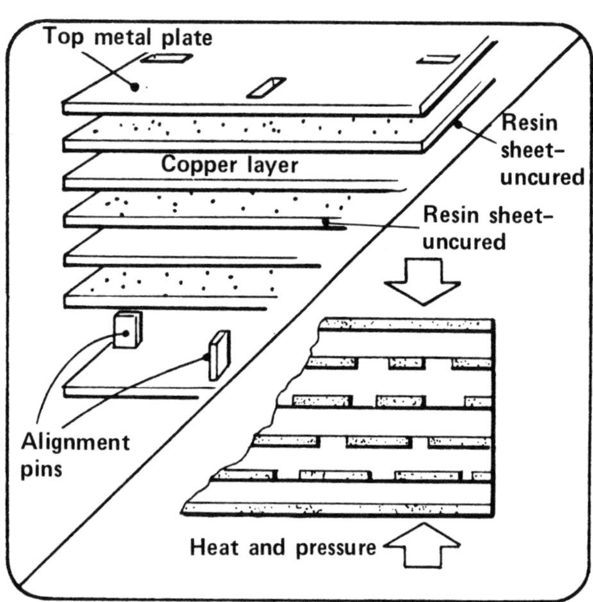

Stage 4

The layers are fitted in the specified order into a fixture consisting of two metal plates with accurate alignment between them. This "sandwich" is then placed in a press and heated under pressure.

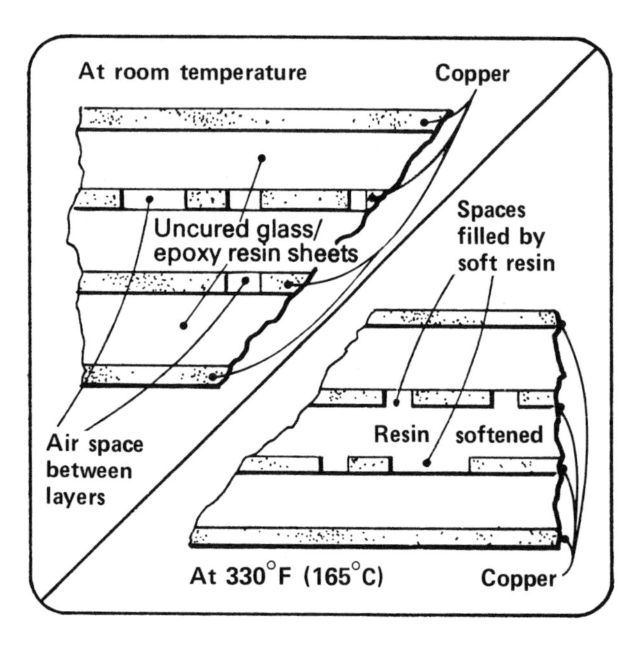

At temperatures approaching 330° F (165° C) the resin in the uncured sheets starts to soften. It then runs under pressure to fill the air spaces between the layers.

At 330° F (165° C) the resin starts to cure and eventually becomes hard and is bonded to the adjoining layers. All insulation sheets are cured when the resin has hardened.

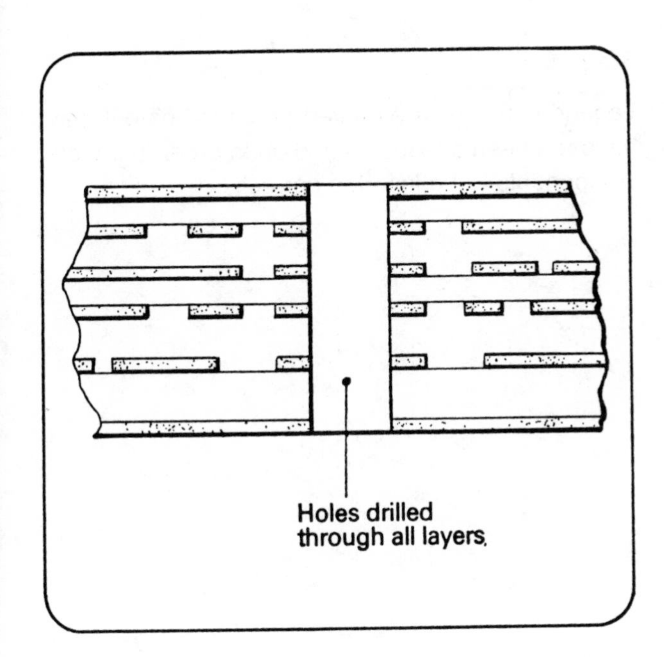

Stage 5

The board is removed from the press and fitted onto a jig. Holes are then drilled in it in order to enable connections to be made between the copper layers at the next stage.

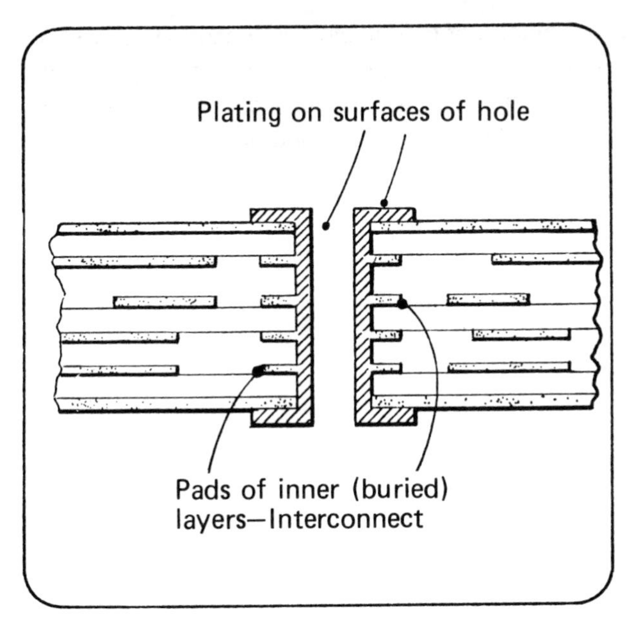

Stage 6

The holes that have been drilled are then "plated-through" with copper by the "through-hole" plating method previously described.

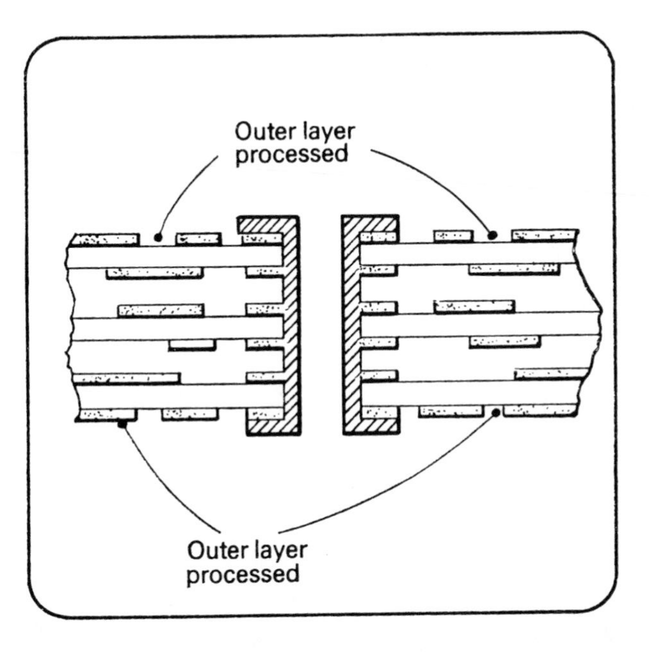

Stage 7

The top and bottom surfaces are then treated in the same manner as a double-sided plated-through-hole board.

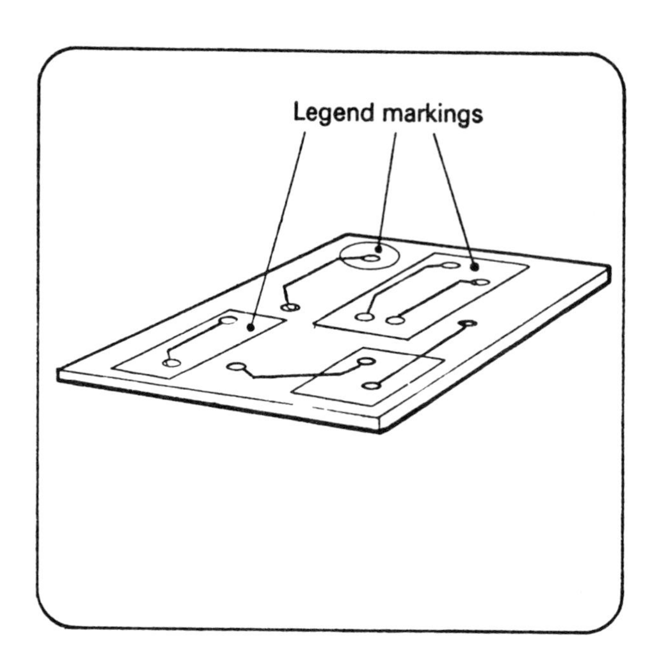

Stage 8

Legend markings are printed onto the board if they have not been already etched on in order to enable components to be inserted correctly.

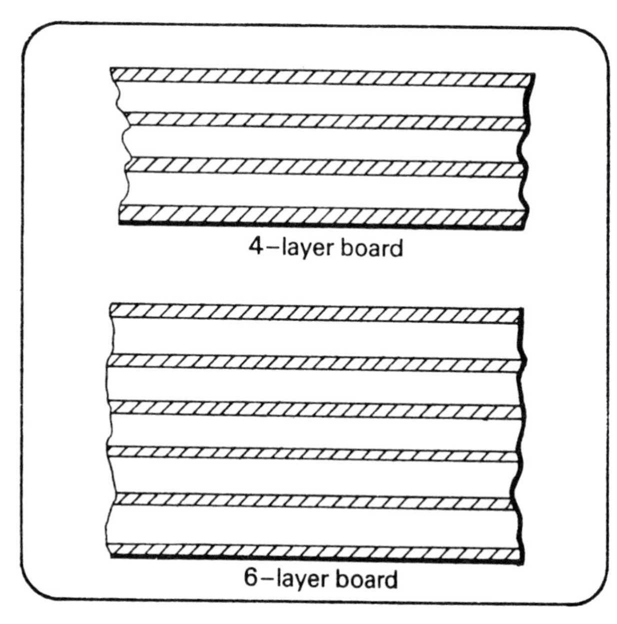

The method of manufacturing described above has produced a board with an even number of copper layers. The "number of layers" of a multilayer board refers to the number of copper layers present.

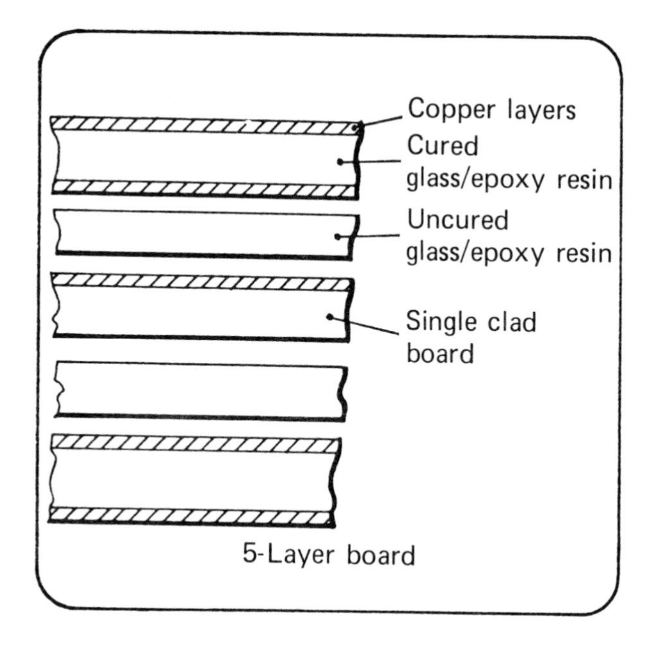

To produce a multilayer board with an odd number of layers, one of the boards must be single-clad, that is, copper foil on one side only.

8″ X 6″ (200 X 150 mm)
one of a batch of 50 6 layers

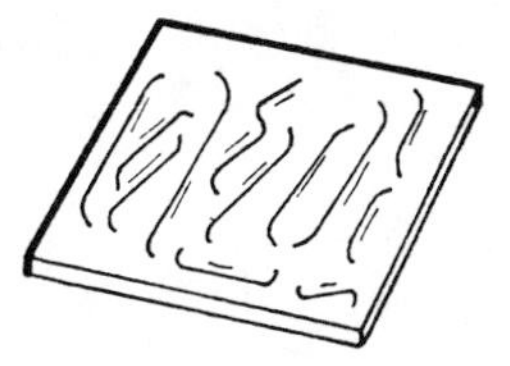

Approximately 10 X the price of an ordinary printed circuit board

8″ X 6″ (200 X 150 mm)
One off only 6 layers

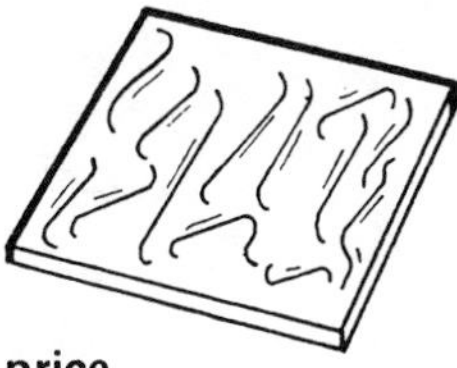

Approximately 20 X the price of an ordinary printed circuit board

12″ X 10″ (300 X 250 mm)
Development board only one of its kind 10 layers

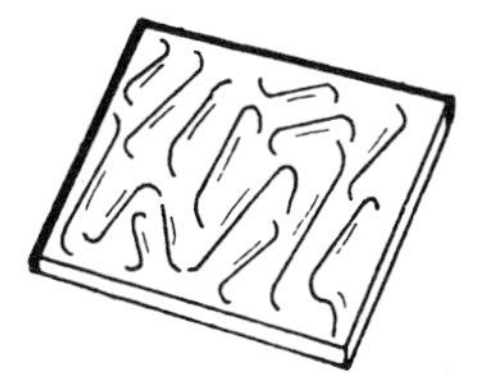

Up to 50 X the price of an ordinary printed circuit board

Special precautions in handling

The cost of multilayer boards depends on a variety of things such as the number of layers, the size of each board, and the number to be made. However, the value of one board may be anywhere between 10 and 50 times the value of an ordinary printed circuit board. Also relevant is the fact that many of the faults that can occur *cannot be repaired.*

VERY GREAT CARE MUST BE TAKEN WITH THE BOARDS.

All the precautions that apply to ordinary printed circuit boards must be taken when handling and using multilayer boards.

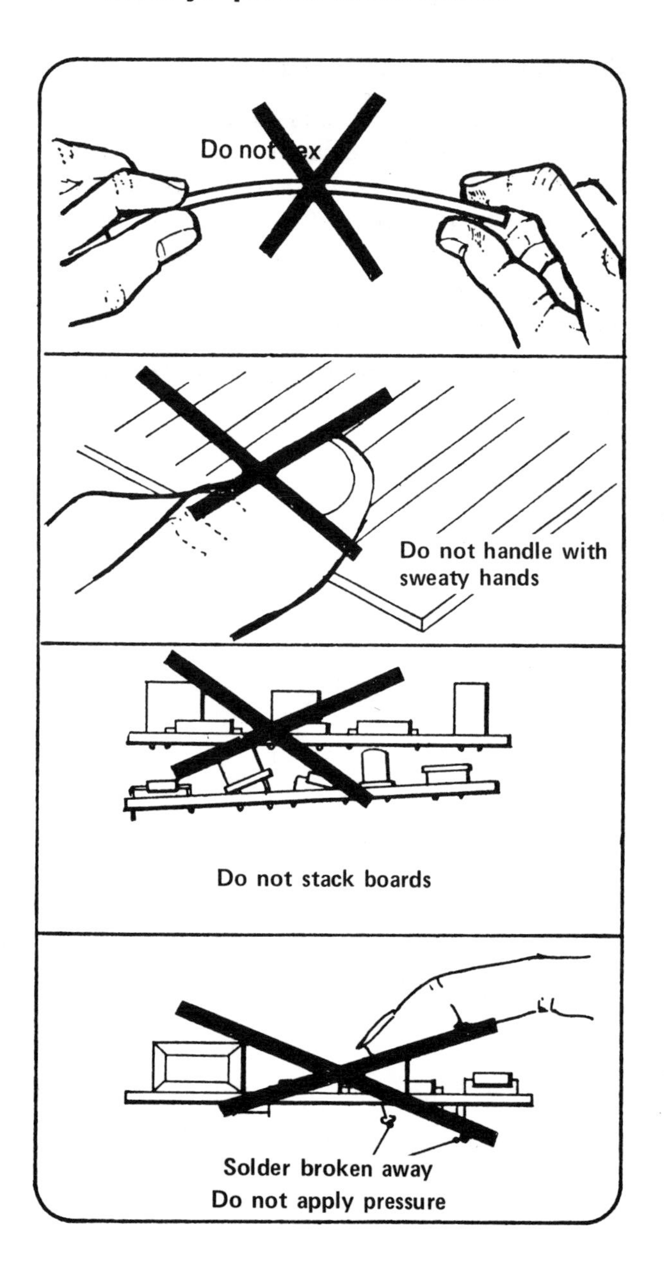

DO NOT HANDLE WITH SWEATY HANDS.

DO NOT STACK BOARDS.

DO NOT APPLY PRESSURE TO COMPONENTS AFTER SOLDERING.

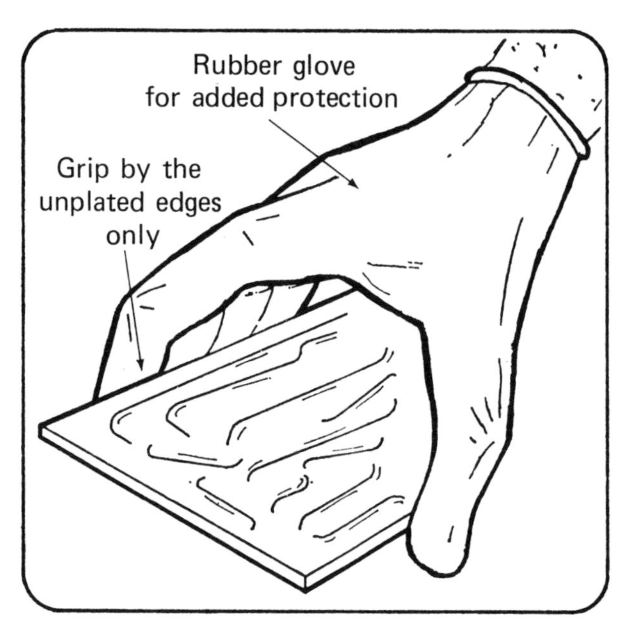

The correct method of handling the boards is to grip them by the unplated edges.

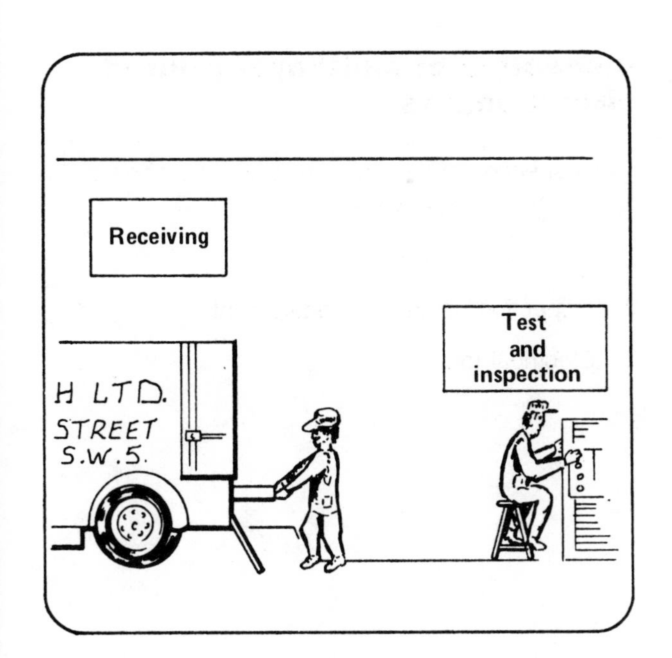

Test and inspection of materials

To minimize the risk of failure, every sheet of double copper-clad board is carefully inspected and tested before it is used for the manufacture of a multilayer board.

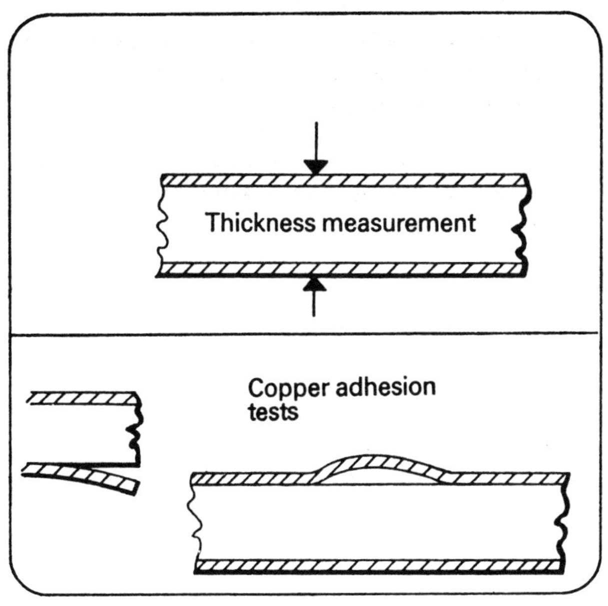

Two of the properties that are tested are thickness and adhesion of copper. The sheets are also carefully inspected for surface flaws.

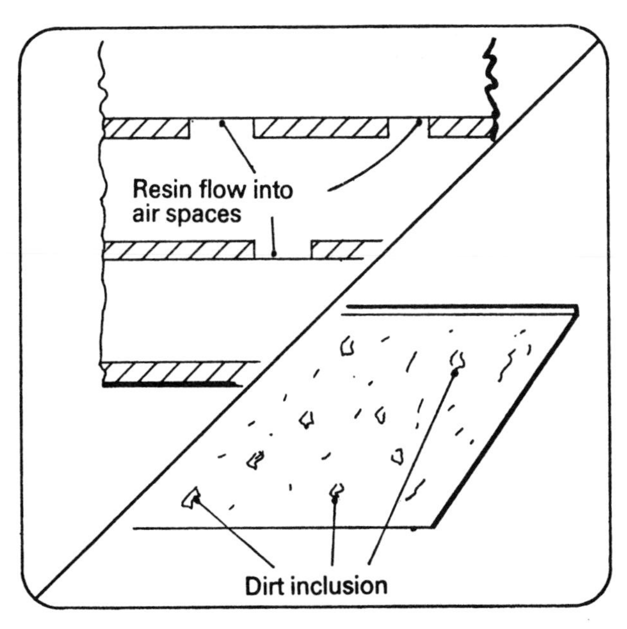

The sheets of uncured insulation are tested for curing time, resin content, resin flow, and any dirt inclusions.

Inspection of multilayer printed circuit boards

The inspection of multilayer printed circuit boards is carried out in three ways.

1 Visual inspection.

2 Inspection using x-ray equipment.

3 Electrical testing.

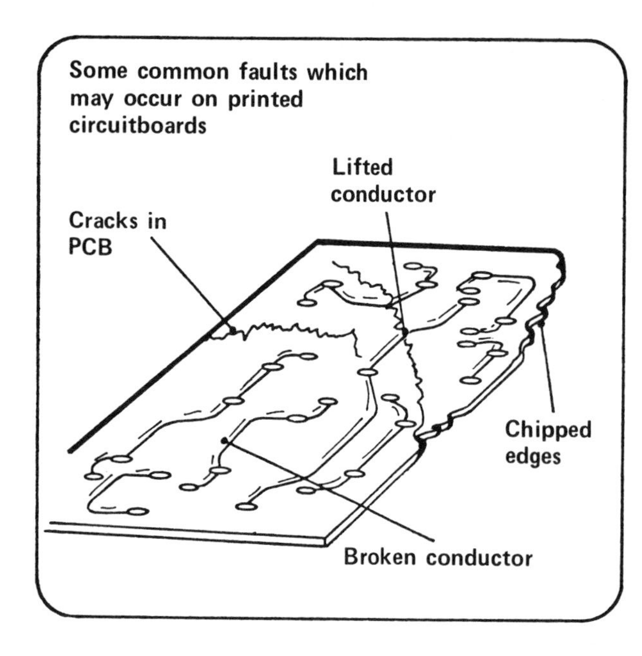

1 Visual inspection. The board is inspected for any of the faults that may occur on an ordinary printed circuit board plus

(a) Microscopic examination of the cross-section of off-cuts to insure that bonding curing has occurred.

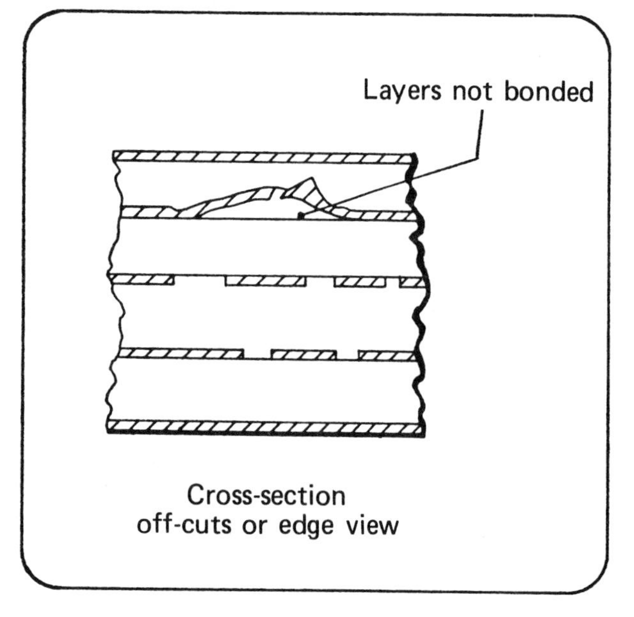

Cross-section off-cuts or edge view

(b) Checking for warping and corrugation.

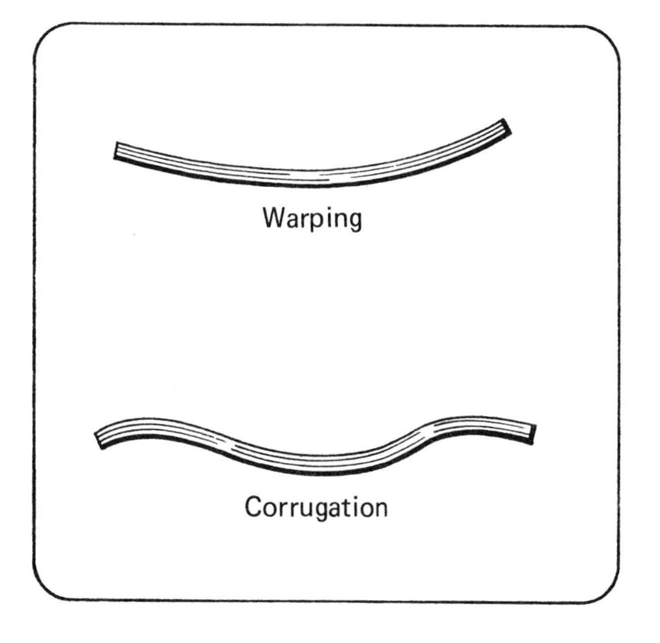

2 Inspection using x-ray equipment.

(a) Alignment of internal pads in relation to the holes.

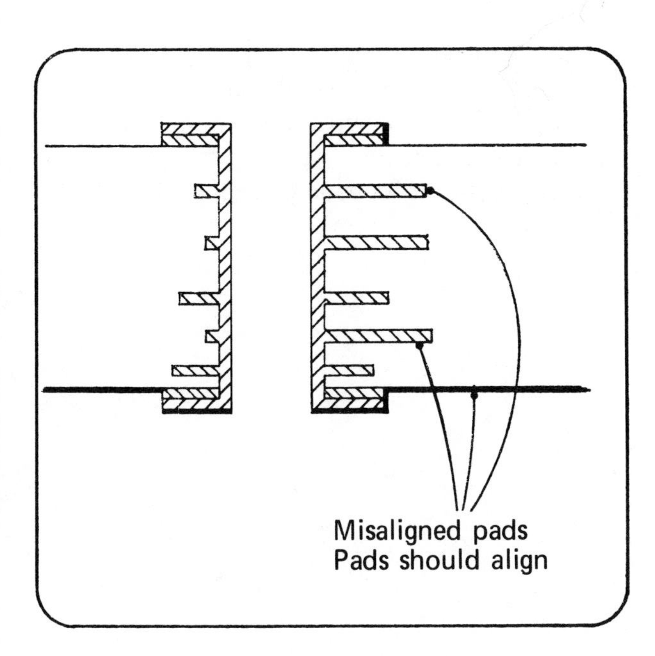

(b) Internal wiring errors or failures.

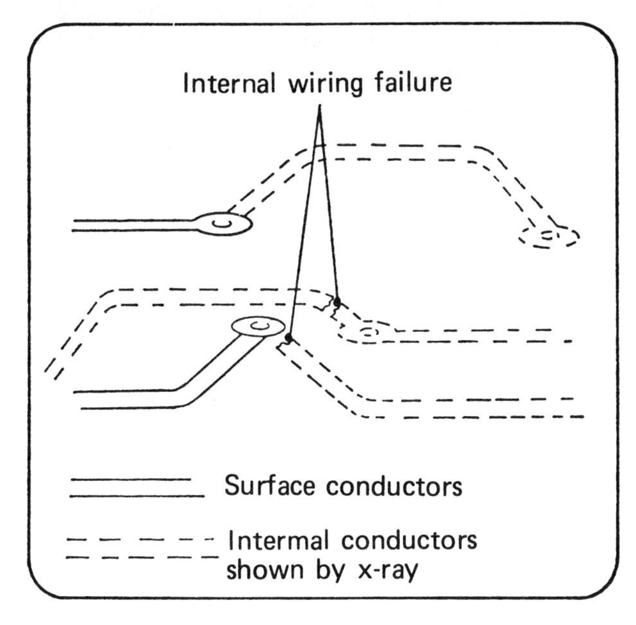

(c) Checking for dirt or air spaces between the layers.

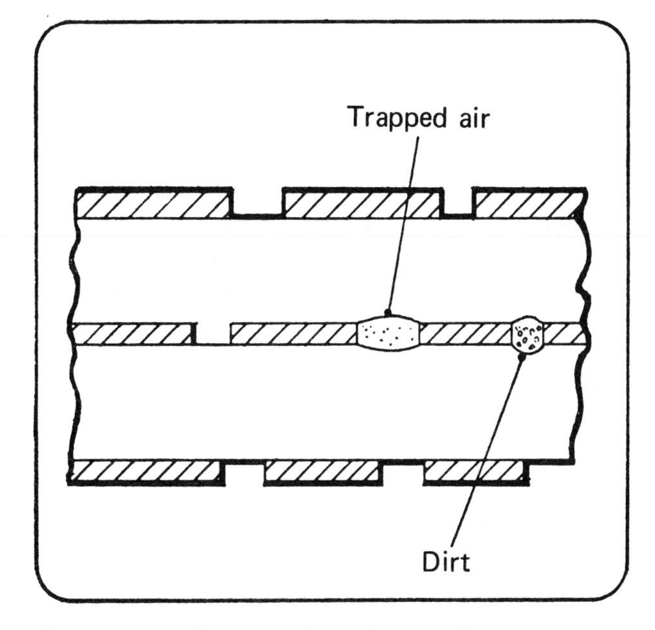

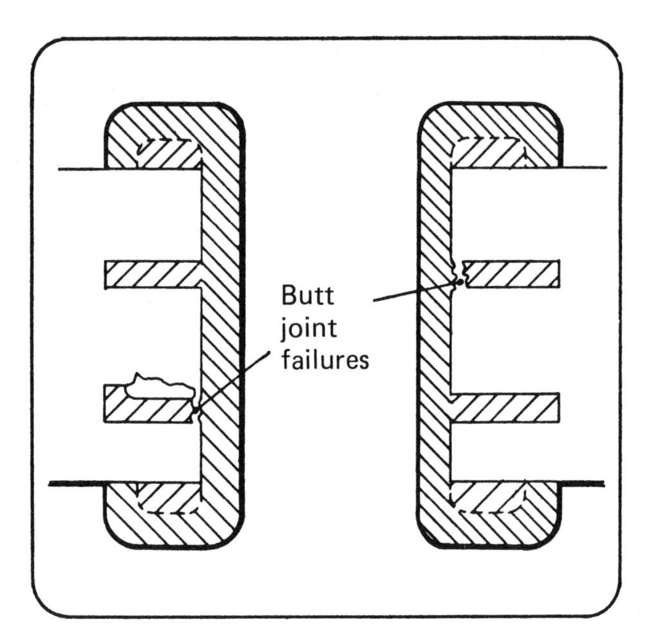

(d) Examination for butt–joint failures that are often caused by overheating during manufacture, component assembly, or repair. In this type of fault, the internal pads are broken away from the copper plating on the inside surfaces of the hole.

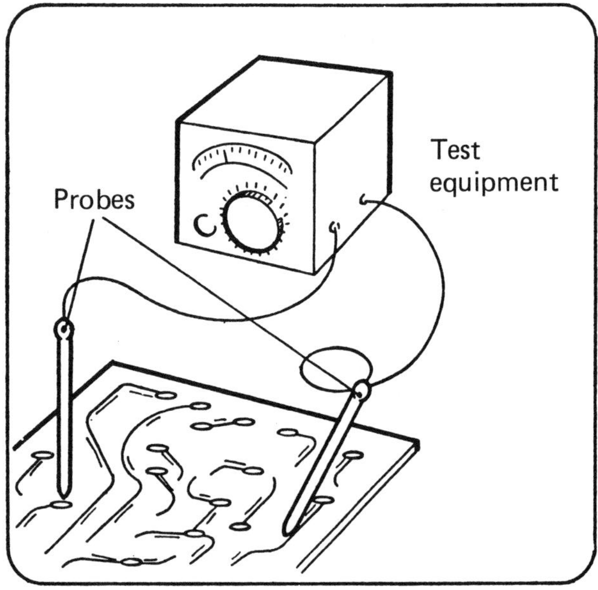

3 Electrical tests. Using appropriate types of test equipment to check for circuit continuity and breakdown of the insulating materials.

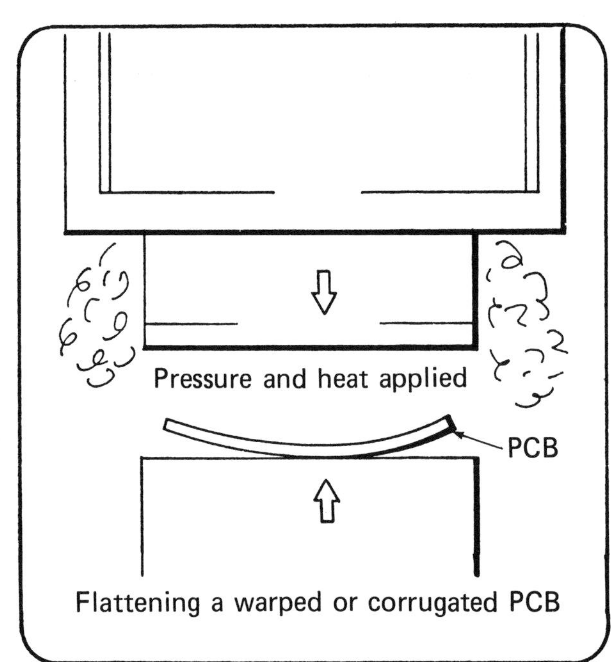

Flattening a warped or corrugated PCB

The only faulty boards that can be easily repaired are those that are warped or corrugated and possibly those with wiring errors on the outer surfaces.

Quality

Other faults may occur that cannot be remedied and in such cases the board must be rejected. Therefore, great care must be taken in all stages of manufacture and assembly.

Test items

1 List three advantages of multilayer printed circuit boards.

(a) ______________________________

(b) ______________________________

(c) ______________________________

(See page 55.)

2 To correctly position the double copper-clad boards when they are joined together, the technician makes accurate _______ holes in each board.

(See page 57.)

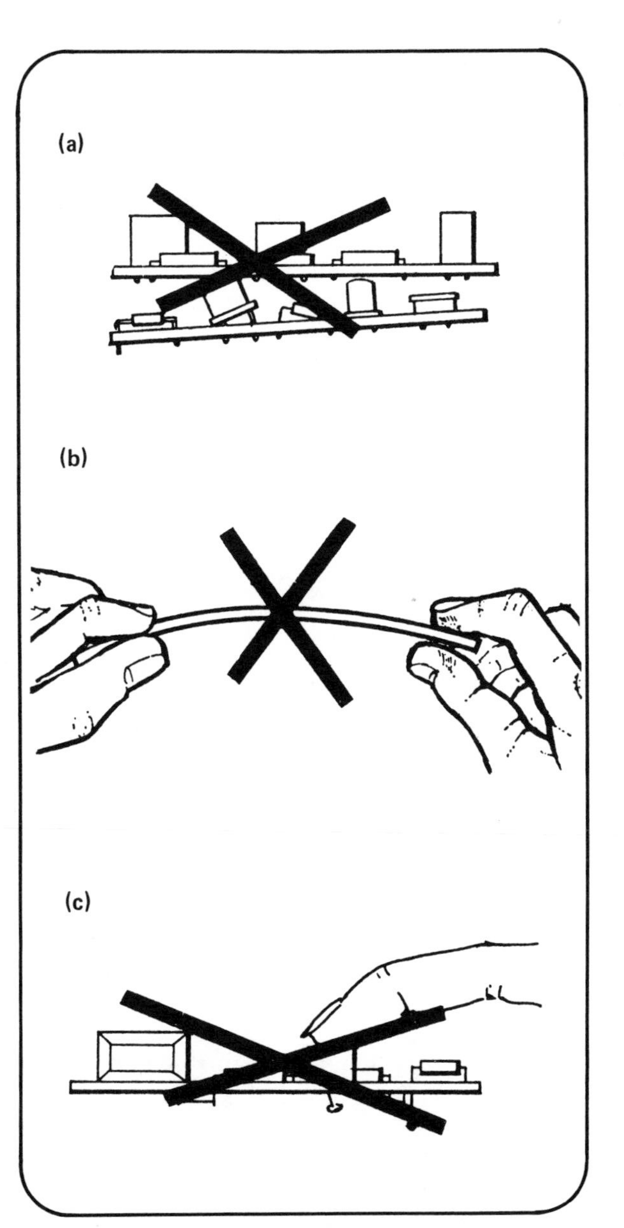

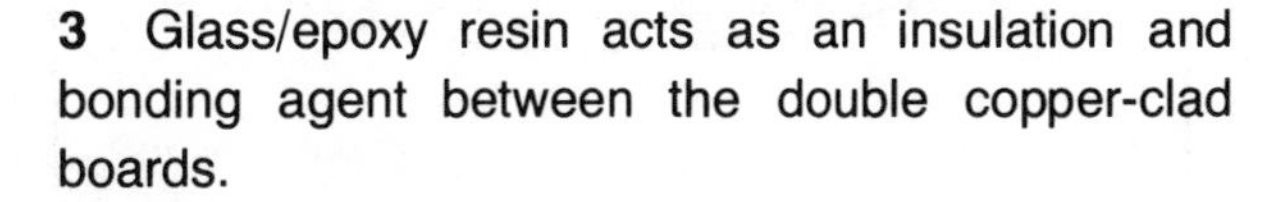

3 Glass/epoxy resin acts as an insulation and bonding agent between the double copper-clad boards.

True or false? (Circle one.)

(See page 58.)

4 Which type of board is the most expensive to manufacture?

(a) Double-sided board.

(b) A four-layer board.

(See page 61.)

5 For each of the following diagrams, label the incorrect handling procedure.

____

(See page 62.)

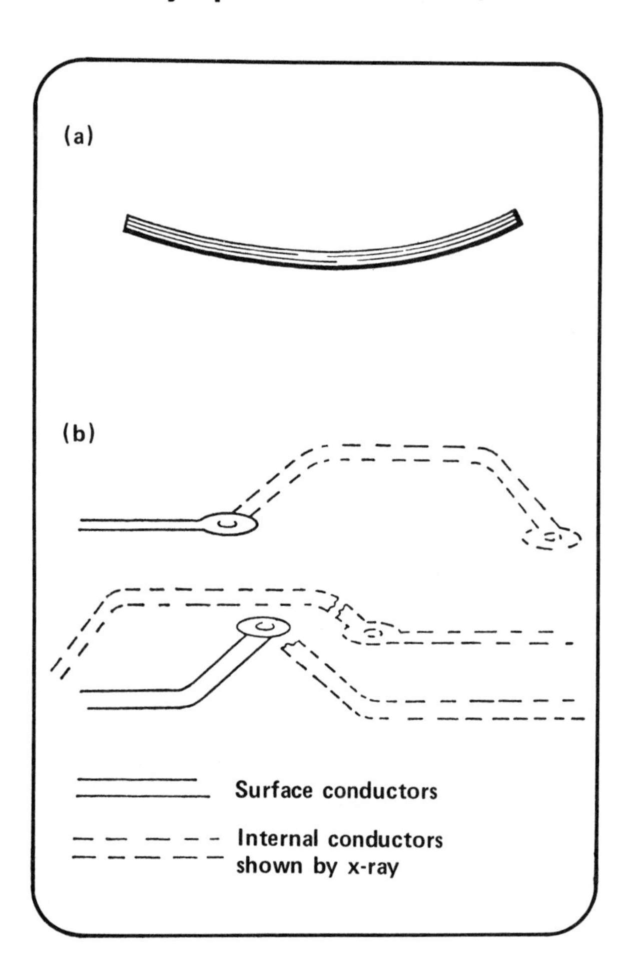

6 Label the type of fault in the diagram.

(See page 64.)

Module six
Microelectronic devices

Introduction

This module examines the basic manufacturing process required to make a microelectronic device. The major manufacturing steps, a discussion of the need for high quality workmanship, and the safety aspects associated with the manufacture of the devices are discussed. Packing of the individual circuits and environmental and electrical testing of the final device are also discussed.

Key technical words

- **Wafer** A thin slice of semiconductor material.
- **Die** A single circuit.
- **Epitaxial** The growth of a single crystal semiconductor film on a single crystal substrate, that is, silicon oxide grown on silicon.
- **Diffusion** A process used to produce a special semiconductor material (*N* or *P*) by adding impurities to a substrate.

Performance objectives

Upon completion of this module, you should be able to:

- Rank the steps required to manufacture a microelectronic circuit.
- List the safety equipment required to manufacture microelectronic circuits.
- List the type of test required before the device is sent to the customer.

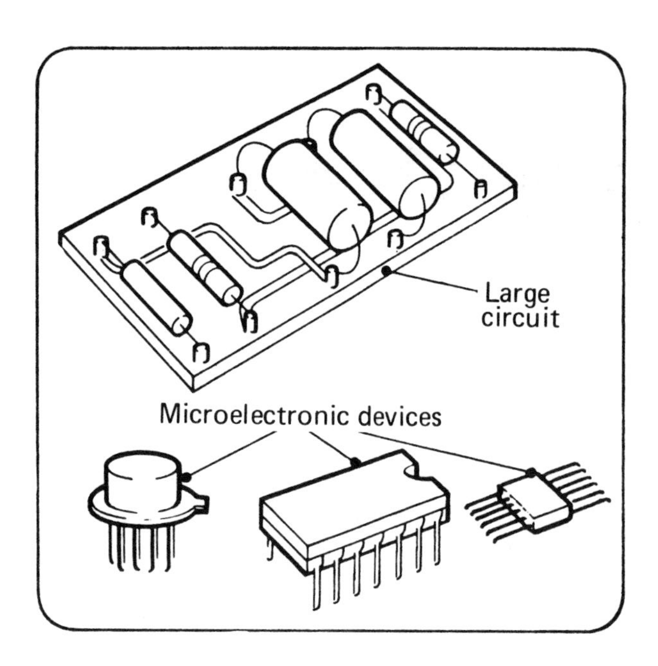

Microelectronic devices

Important

The production of a microelectronic device involves a series of highly skilled and sometimes highly complicated processes. The cost of a device is between a few cents and several dollars.

A microelectronic device can replace a large circuit board assembly containing numerous components and a complex network of wire; thus, there is a large saving in mass and space. The illustration shows one example where the large circuit board may be replaced with one device.

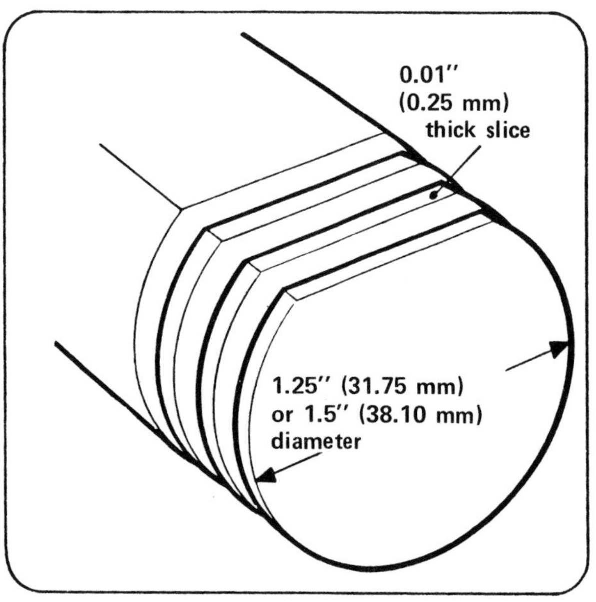

Method of production

Microcircuits are etched on silicon slices or wafers that are cut from a bar.

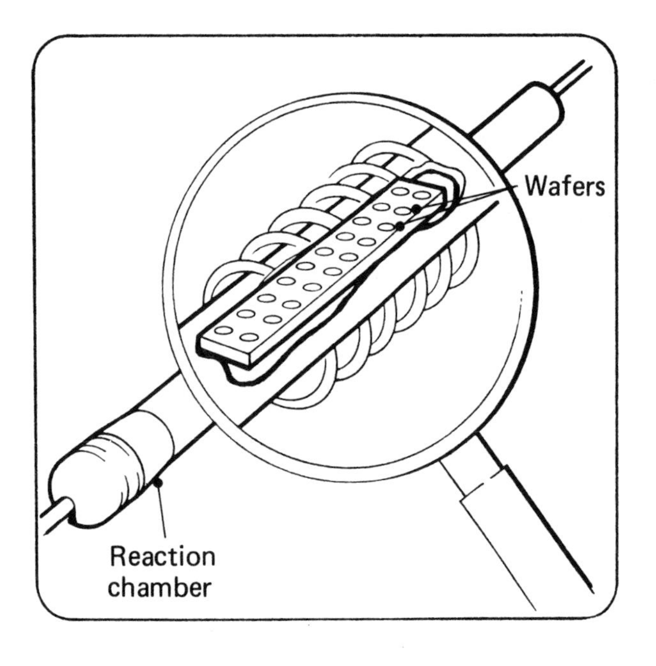

Epitaxial growth

The slices are placed in a reaction chamber to undergo an epitaxial process, that is, one face of the slice is coated with silicon oxide.

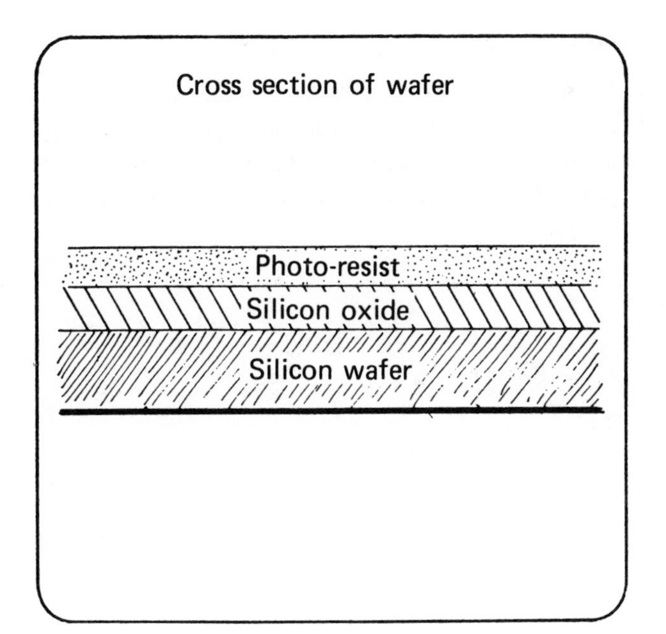

Photo resist

The wafers are coated with a photo resist.

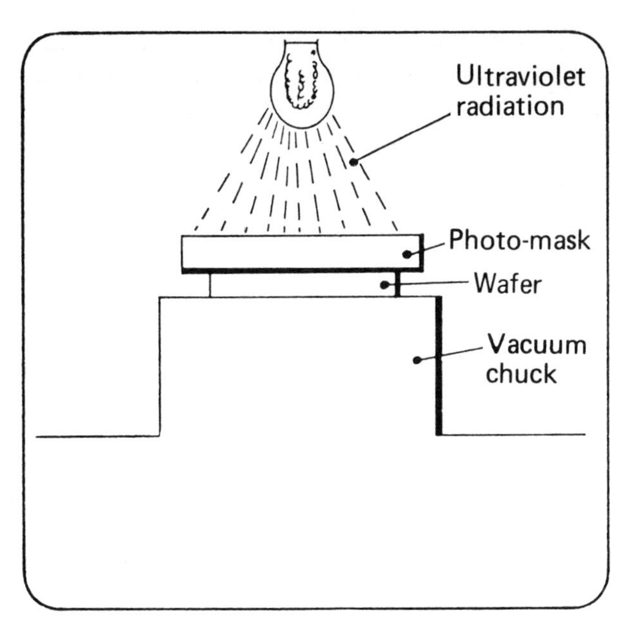

Alignment

The wafer is aligned under a photomask that contains a pattern of part of the required circuit. The pattern is repeated approximately 400 times. The wafer will contain approximately 400 identical but separate circuits each 0.025 in. (0.635 mm) to 0.0625 in. (1.588 mm) square depending on what type of circuit is being produced. The wafer is exposed to ultraviolet radiation through the mask.

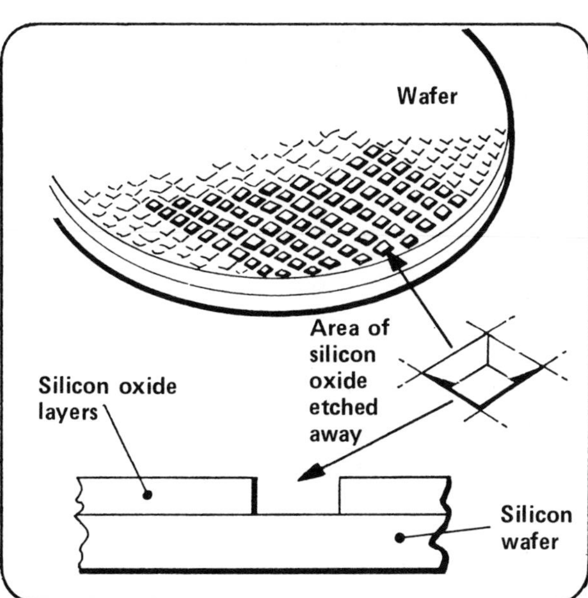

Developing and etching

After the wafer is exposed, it goes through a developing process and then a chemical etching process.

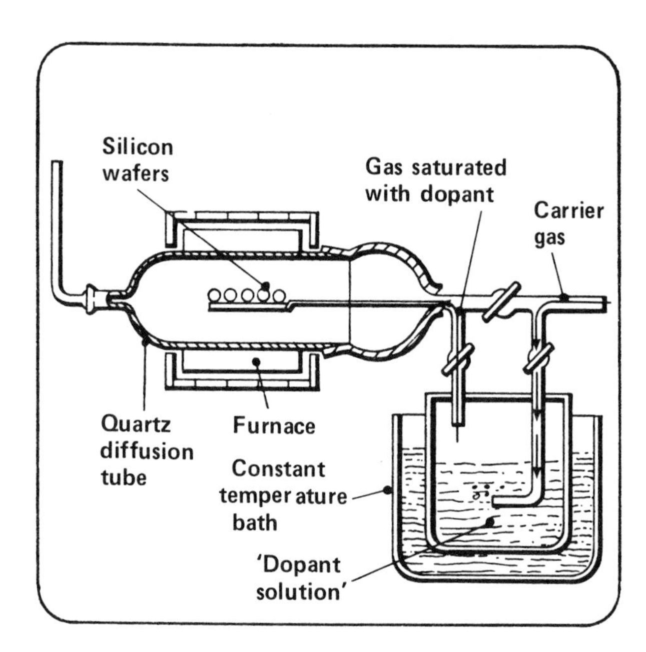

Diffusion

The slice undergoes a diffusion process. A carrier gas passes through a solution high in impurities called a "dopant." A gas saturated with dopant is taken off the top of the solution and goes into the diffusion tube.

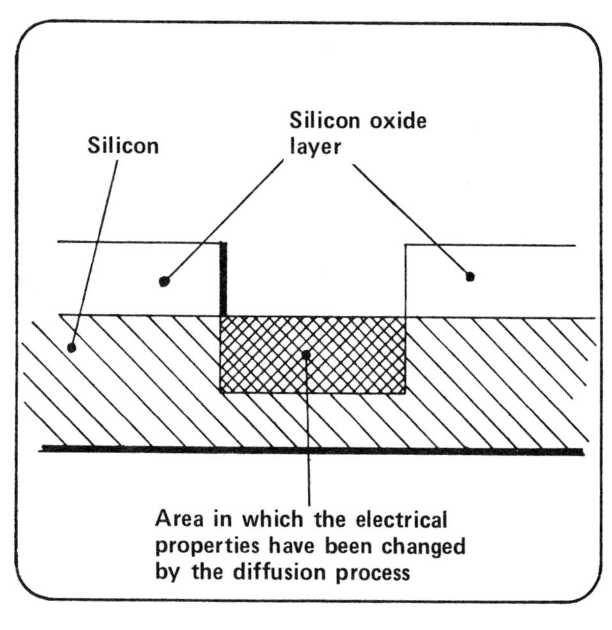

Area in which the electrical properties have been changed by the diffusion process

This process has the effect of changing the electrical properties of the silicon slice at the places where the silicon oxide has been etched. An example is the making of a *N*-type or *P*-type region required for a diode or transistor.

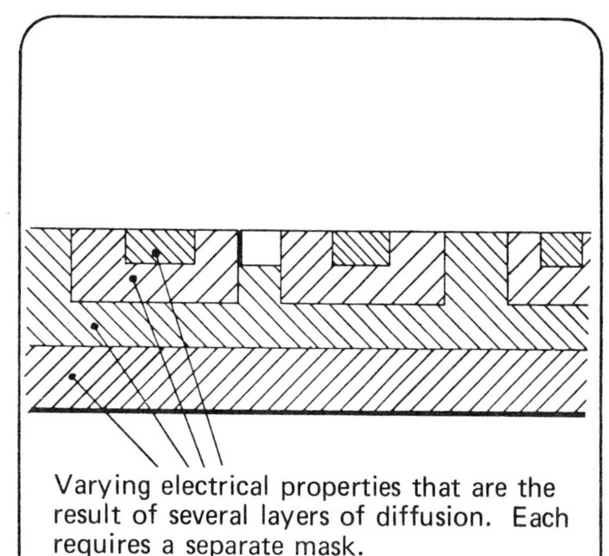

Varying electrical properties that are the result of several layers of diffusion. Each requires a separate mask.

Following epitaxial growth, the series of processes, photoresist, alignment, development, etch, and diffusion, are repeated several times until there is a complex system of areas of varying electrical properties. The areas have the properties of components, for example, resistors, diodes, and transistors. A diode has a *N*-type and a *P*-type region, one on top of the other.

> ***Note:*** A different mask is used each time in the diffusion process. Each mask contains a different part of the circuit.

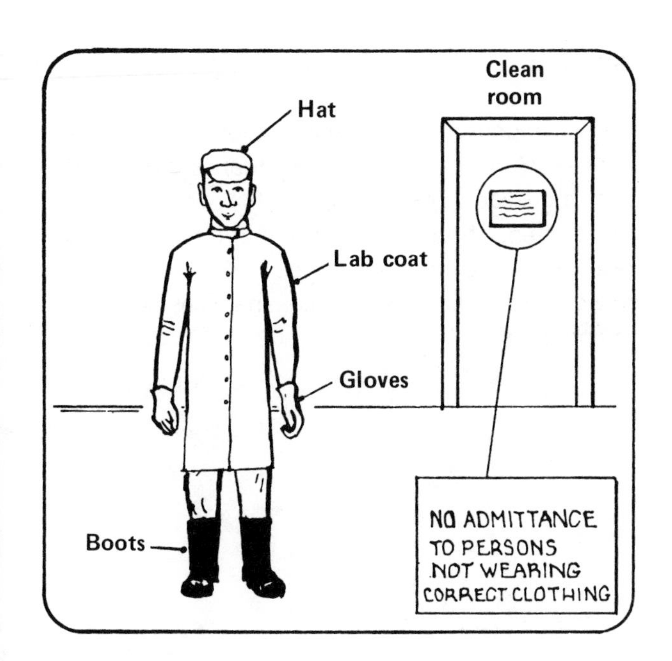

Quality

Photo resist, alignment, development, and etch processes are carried out in environmentally controlled dust-free areas. Special clothing must be worn to protect the wafer from foreign particles, that is, dust, hair, and loose skin.

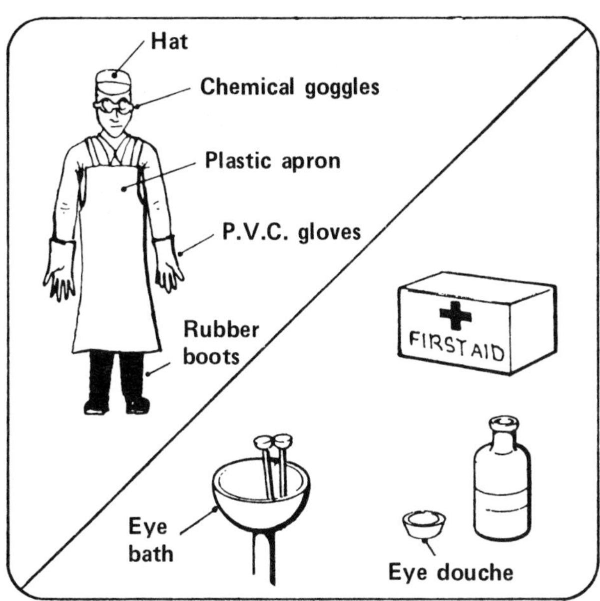

Safety

Considerable use is made of chemicals during the processes. Wear protective clothing when handling reagents and take care to avoid spilling them. Clean up spilled reagents immediately, using the specified materials. For splashes on the skin or in the eyes wash well with cold water or rinse with eye douche and seek first aid.

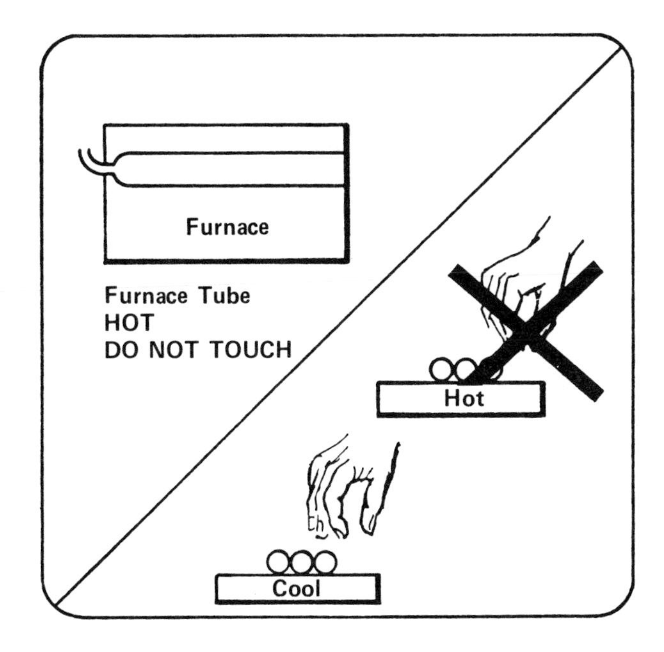

Safety

The furnaces used in the epitaxial and diffusion processes operate at high temperatures. Do not touch the tube in the furnace in which the wafers are placed. Allow the slices to cool after removal from the furnace.

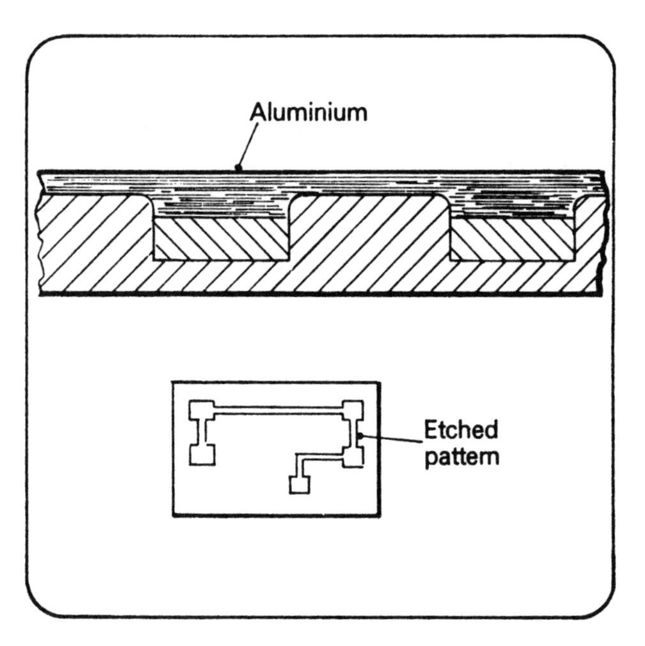

Metallization

The "components" are interconnected by a thin layer of aluminum (0.0001 in.) that is evaporated onto the slice. The pattern of the circuit is then acid-etched. The wafer is heated in a low temperature furnace to produce a good electrical contact between the silicon and the aluminum links between components.

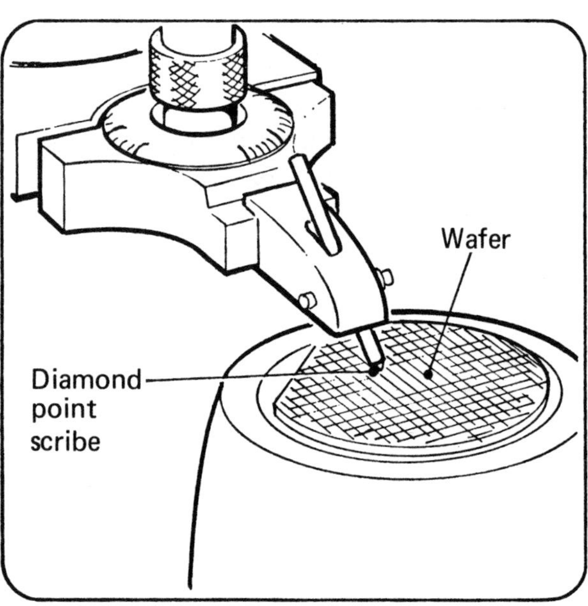

Scribing

A line is scribed between the circuits in preparation for cutting the wafer. This technique is similar to the cutting of glass.

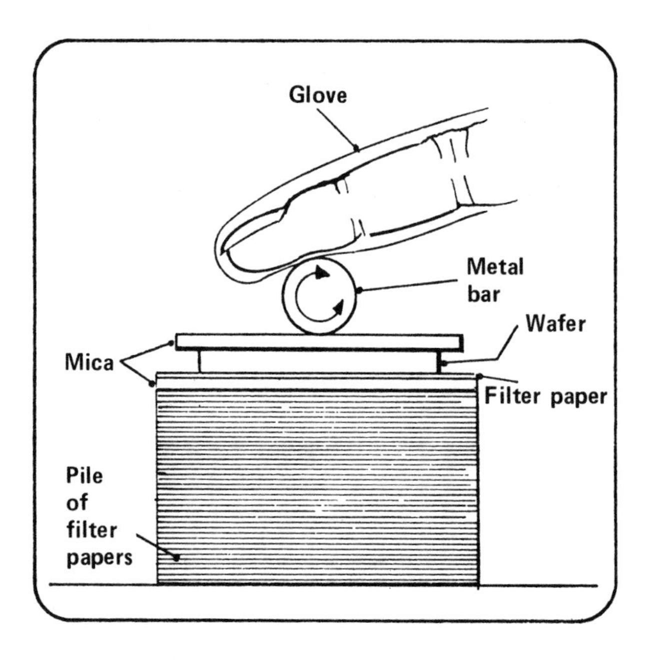

Dicing

The wafer is cut up into individual "chips" or dice. Each chip or die contains a circuit. A metal bar rolls over the wafer to break the chips apart.

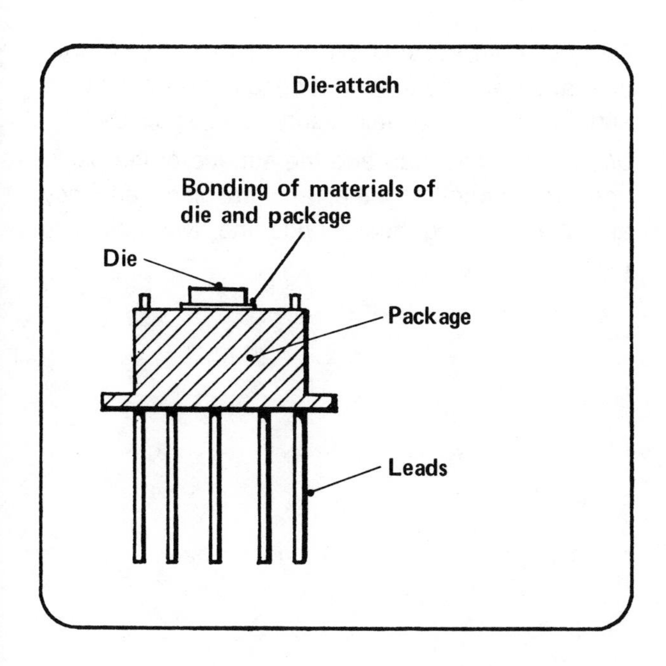

Die-attach

The die is attached to a package by one of three methods.

Eutectic A thin layer of gold is deposited on the back side of the die. The package is heated to a temperature high enough to melt the gold, bonding the die and the package.

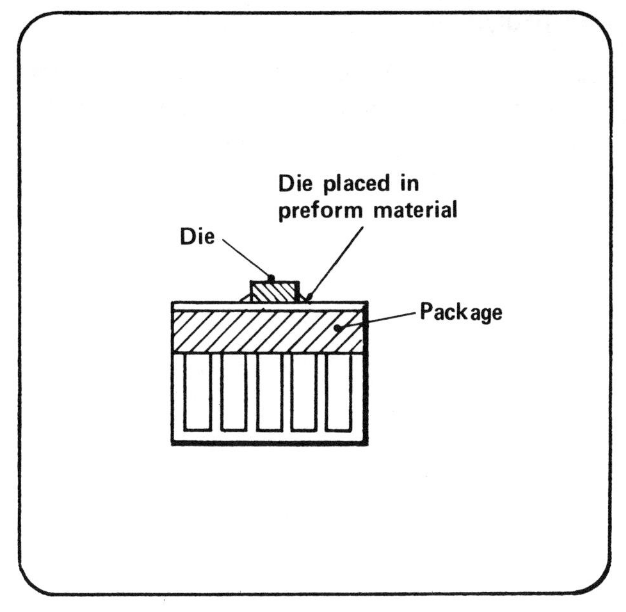

Preform A special material that will adhere to both the die and the package is used. The material is melted on the die-attach area of the package and the die is placed in the material.

Epoxing An epoxy glue is used to attach the die to the package.

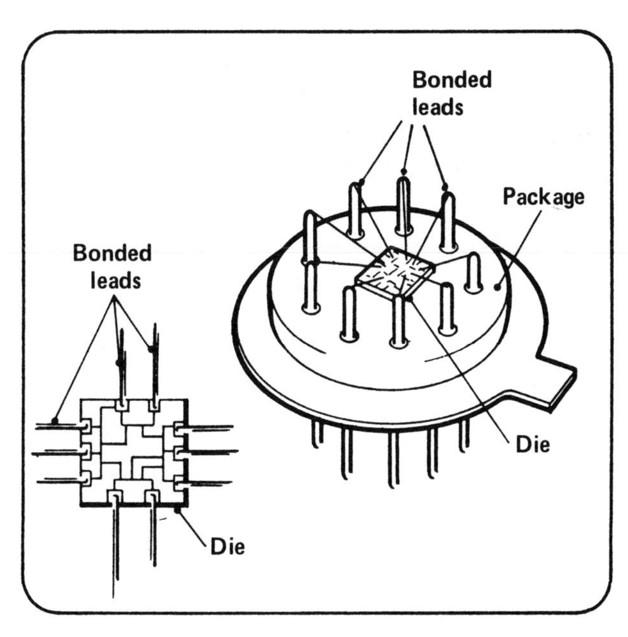

Bonding

Fine leads are bonded to the device. A typical lead diameter is 0.0015 in. (0.0381 mm). The leads are made from aluminum or gold. Two common methods of wire bonding are

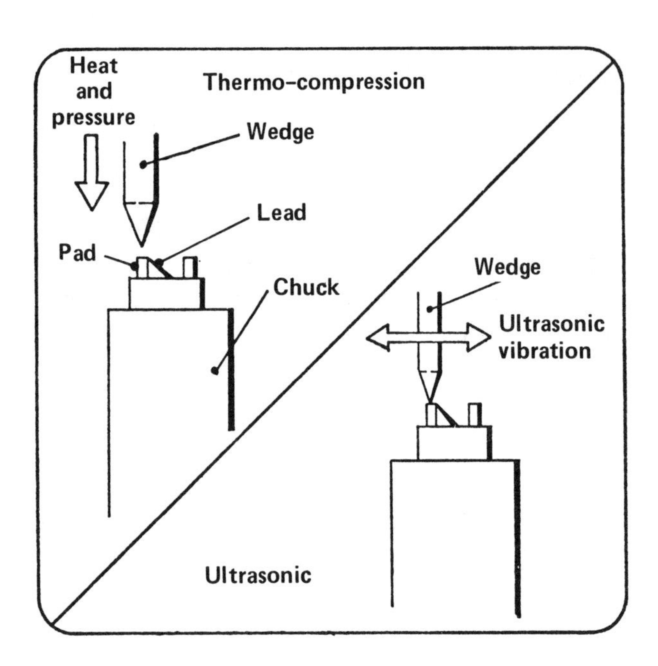

Thermo–compression This method uses both heat and pressure to form a bond between the lead and the pad. It is usually used with gold leads.

Ultrasonic The lead and the surface of the package are vibrated together at ultrasonic frequency and the resulting heat bonds the two materials together.

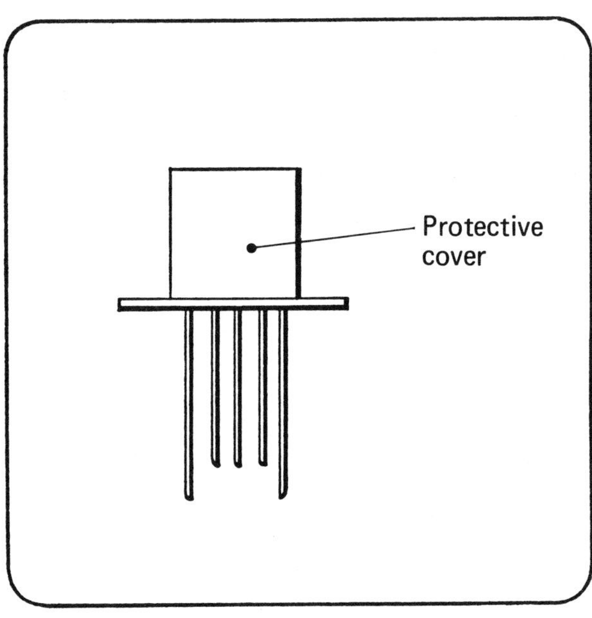

Encapsulation

The fragile die is covered by a protective cover. These covers or packages are made of metal, ceramic, various plastics, and epoxy.

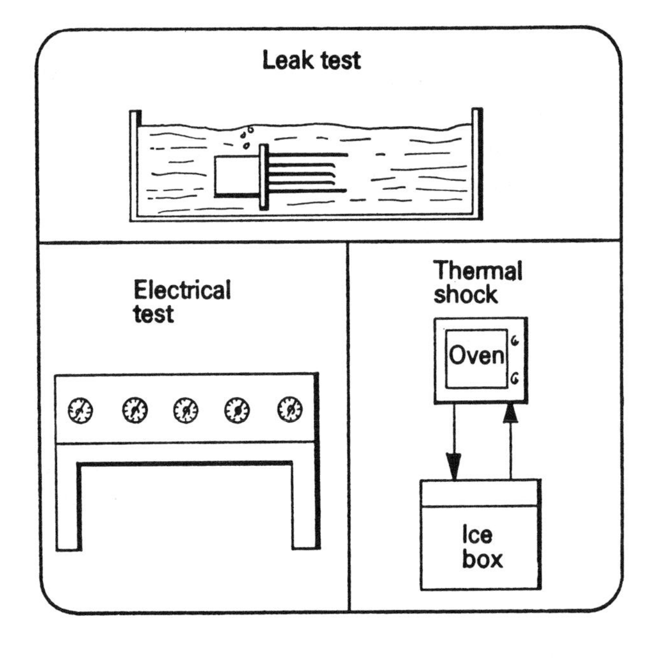

Test

The devices receive various environmental and electrical testing to insure their reliability before being delivered to the customer. The devices may be individual components such as diodes or transistors or they may be complex integrated circuits (ICs).

Test items

1 Choose from the list at the right the missing steps in the manufacturing process given the list of steps on the left.

1 Epitaxial growth	
2 ________________.	Scribing
3 ________________.	Test
4 Developing and etching	Photo resist
5 Diffusion	Alignment
6 Metallization	Bonding
7 ________________.	
8 Dicing	
9 Die-attach	
10 ________________.	
11 Encapsulating	
12 ________________.	

(See page 70, 74, 76.)

2. Complete the blanks on the drawing of the required clothing to protect the wafers from foreign particles.

(See page 73.)

3 List two types of tests normally performed on the devices before they are delivered to the customer.

(a) ________________________________

(b) ________________________________

(See page 76.)

Module seven
Thin film modules

Introduction

This module examines the applications of thin film devices and the manufacturing processes. The major manufacturing steps are described and illustrated.

Key technical words

- **Thin film** A manufacturing process during which the film is deposited on the substrate in a vacuum chamber.
- **Evaporant** Provides the material, usually gold or aluminum, that is deposited on the substrate.

Performance objectives

Upon completion of this module, you should be able to:

- State the common types of evaporants.
- State the purpose of the three chambers of the vacuum chamber.
- State the proper soldering technique for connecting the base to the substrate.

Thin film modules

Thin film technology is used in the manufacturing of resistors, capacitors, and conducting paths. Thin film technology provides a high degree of control and low tolerance parts necessary for complex circuits. The major disadvantages of thin film devices are cost and lack of flexibility. The process is used only on a limited scale for the manufacture of active components.

The process described here may vary between manufacturers, but the principles of manufacturing is the same.

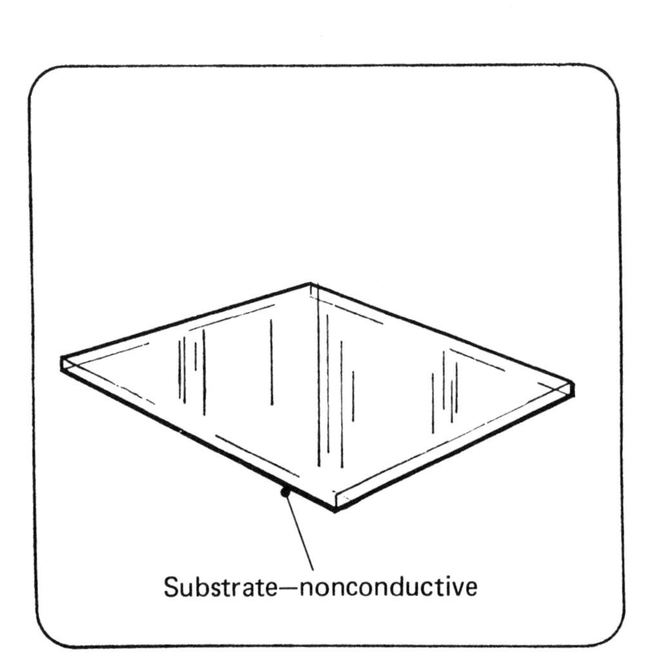

The substrate

The substrate of a thin film module on which a circuit is evaporated, is a nonconductive material such as glass or ceramic. The substrate is coated with a thin film material. A photoresist is applied to the thin film. A photomask is applied to the material that is then exposed, developed, and etched. The photomasked section may vary for each evaporation process.

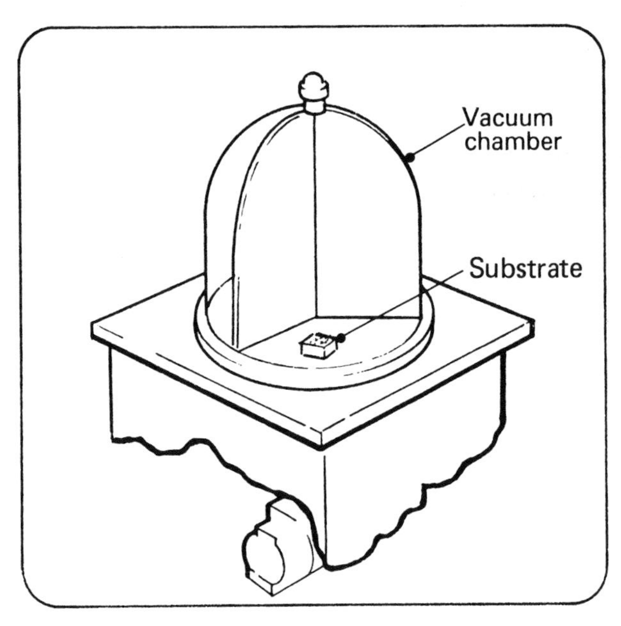

Evaporation equipment

The process of evaporating the circuit is carried out in a vacuum chamber that is divided into three sections.

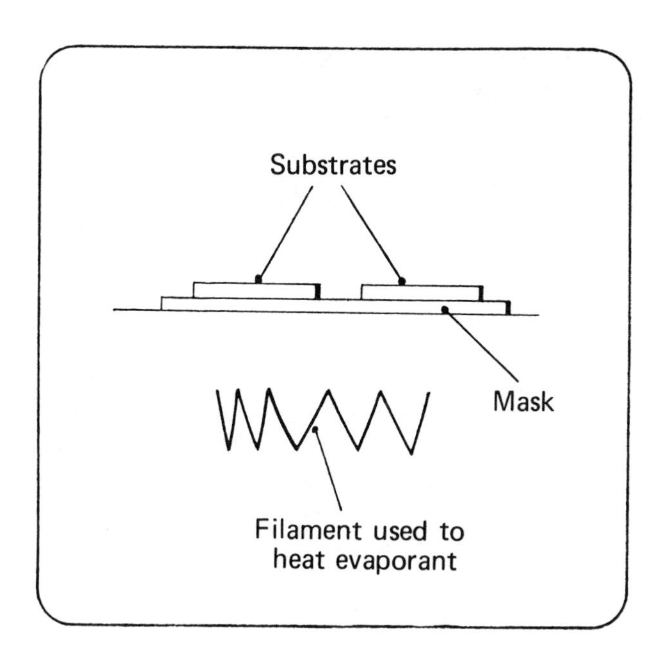

The material to be deposited is heated until atoms of the evaporant "boil" off and condense as a uniform film on the substrate. The evaporant material is a nonrefractory atomic metal such as gold or aluminum.

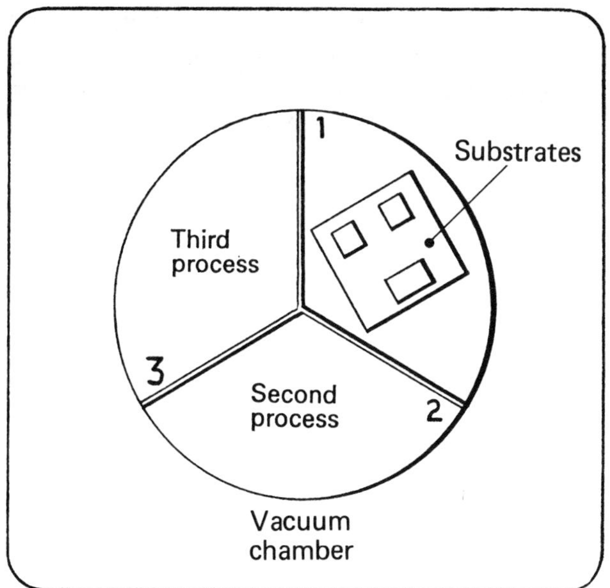

The substrates are placed in section one of the vacuum chamber and are not removed from the chamber until the third evaporation process has been completed in section three.

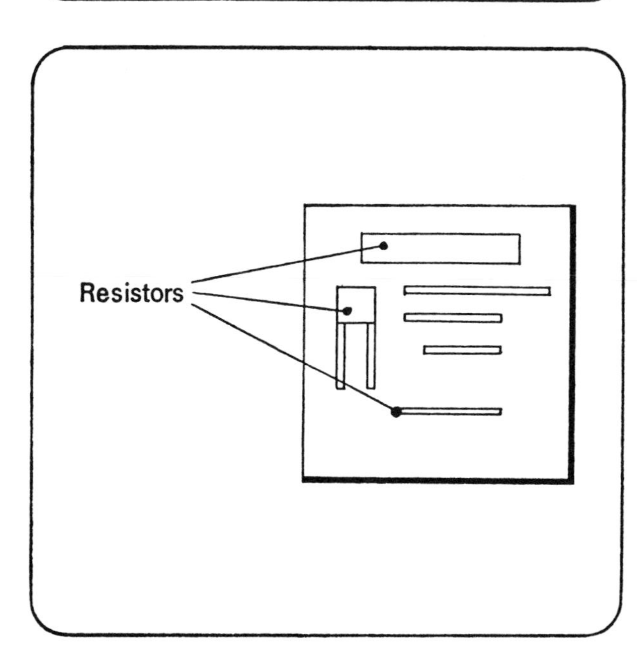

Evaporation process

In section one, the resistors are formed by the evaporant condensing on the substrate in specific shapes and patterns controlled by the photomask.

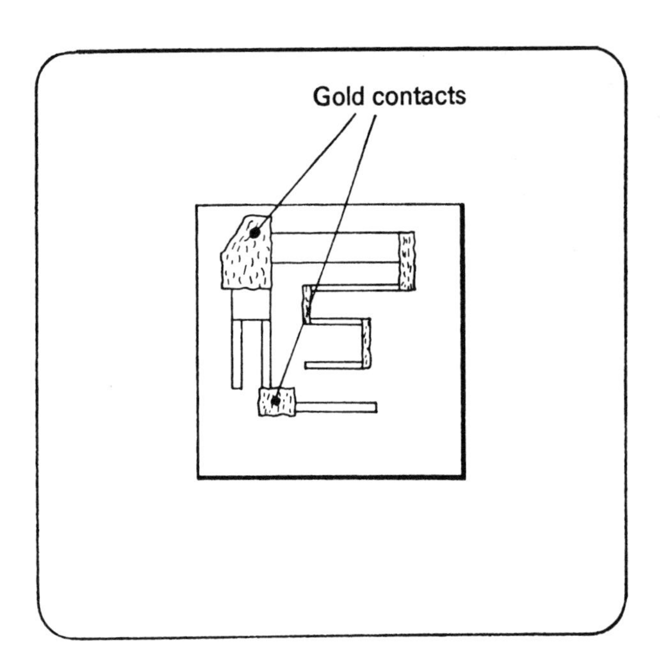

In section two, gold contacts are evaporated onto the substrate.

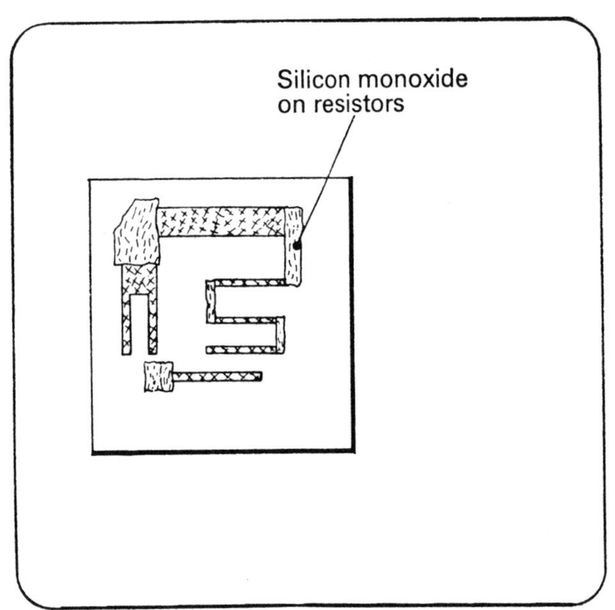

In section three, silicon monoxide is evaporated onto the resistors. This protects the resistors and prevents them from becoming contaminated when they come into contact with the atmosphere.

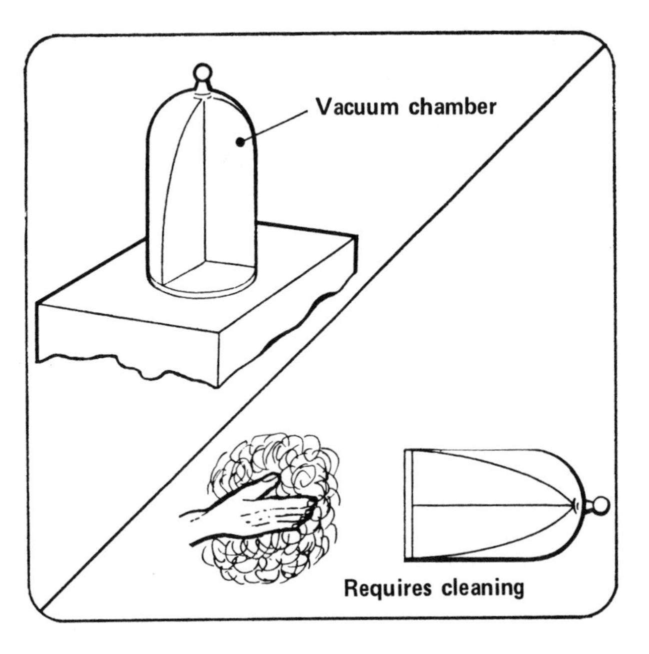

During the evaporation process, the sections of the vacuum chamber become filled with the material being evaporated on the substrate. The inside of the dome becomes coated and has to be cleaned at regular intervals.

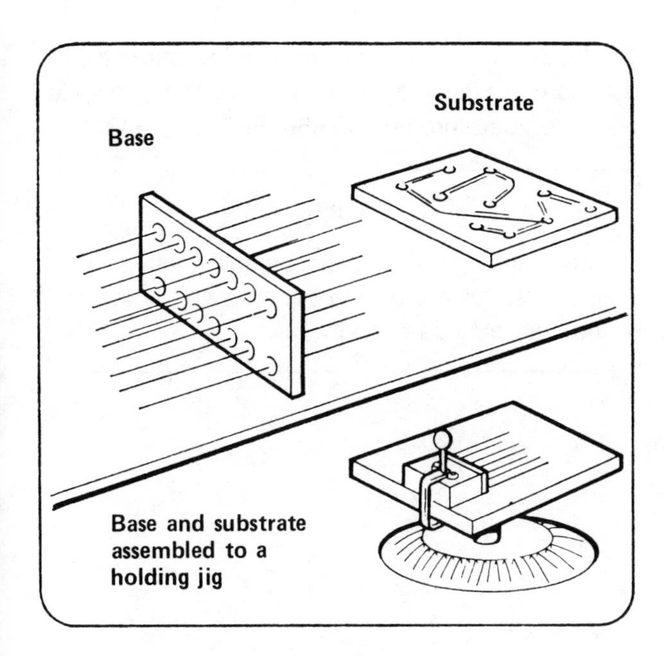

Thin film module assembly

Each substrate is mounted to a ceramic base. Gold wires connect the base leads to the substrate.

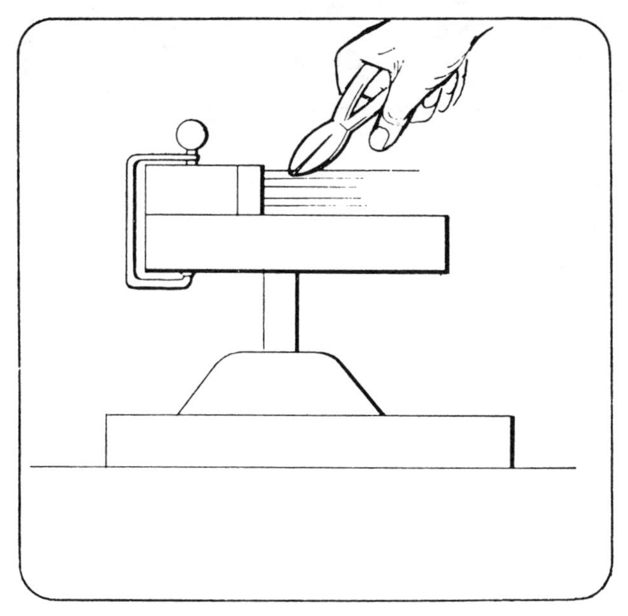

The leads are cut to length.

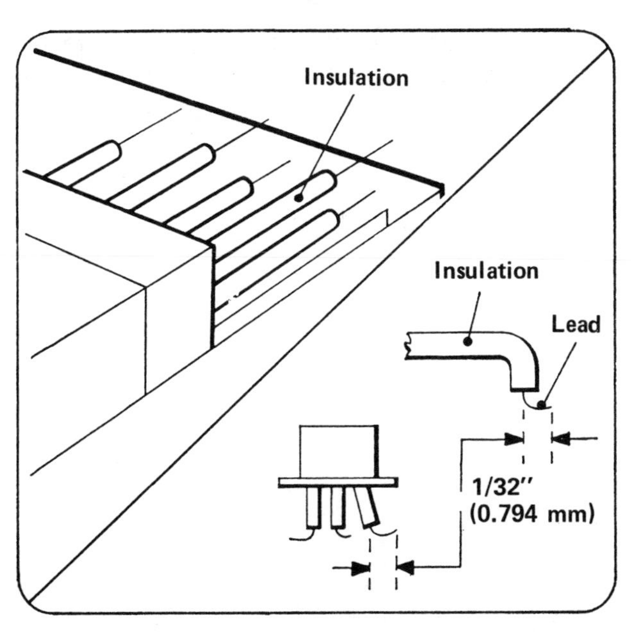

The leads and discrete component leads are insulated and the lead ends are prepared for soldering.

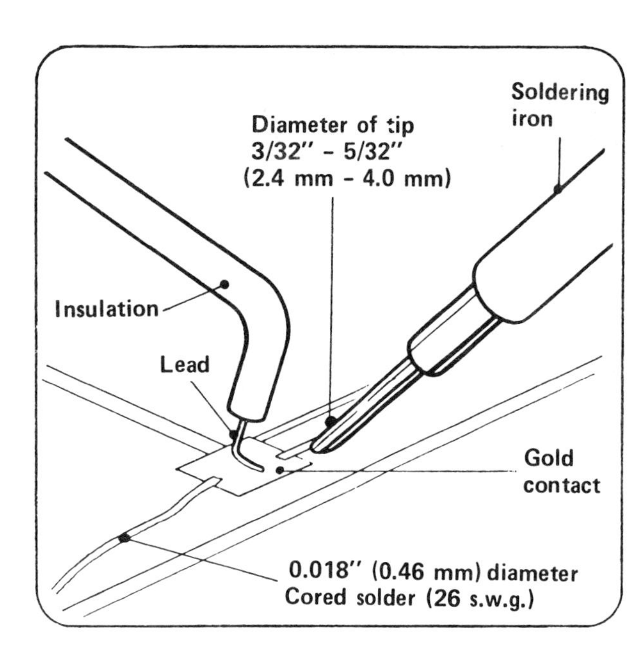

The leads and components are hand-soldered to the gold contacts using a soldering iron with a fine tip at a high temperature 600° to 650° F (315° to 343° C).

> **Quality**
>
> The soldering iron must not be held against the substrate for more than the specified time or the heat may damage it.

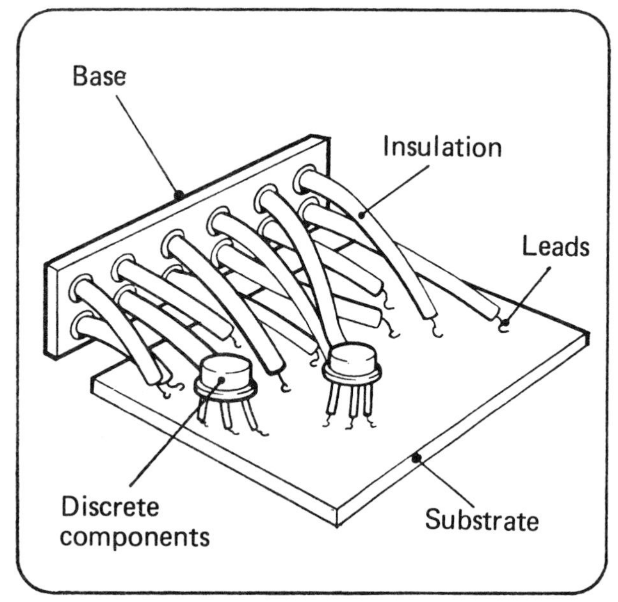

> **Quality**
>
> The leads and components are soldered, starting nearest the base and finishing furthest from it.

Solder the leads from the base to prevent damaging the insulation.

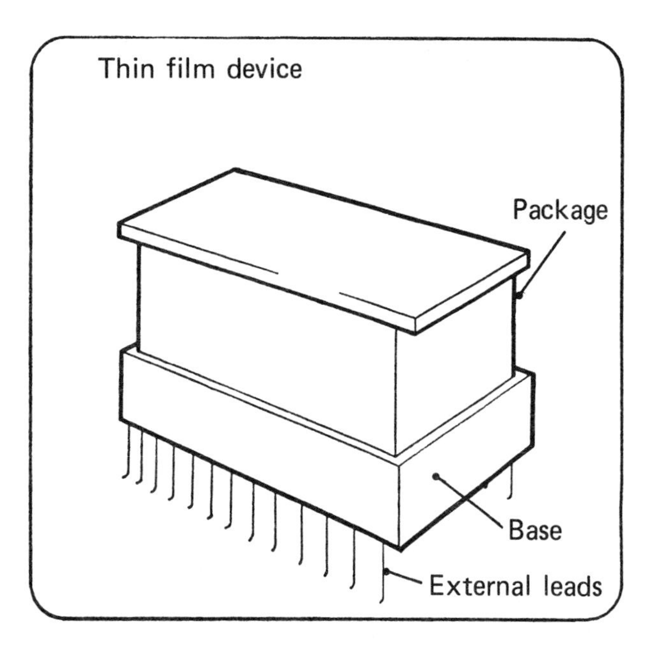

The assembly is fitted into a package that protects the circuit from damage. The thin film module is then complete. The finished component is illustrated.

Test items

1 Name two common evaporant materials.

(a) ______________________________

(b) ______________________________

(See page 82.)

2 In which chamber of the vacuum chamber are the gold contacts evaporated onto the substrate?

____

(See page 82.)

3 When soldering the leads, should you start with the leads nearest the base or those furthest from the base?

____

(See page 84.)

Module eight

Assembling microelectronic devices to printed circuit boards

Introduction

This module examines the basic tools and procedures required to mount microelectronic devices onto a printed circuit board. The assembly of a printed circuit board requires that the right tools be used in each process. The tools must be handled properly and maintained in good condition.

Microelectronic devices normally can be mounted correctly on the board in only one direction. Different keying methods of alignment procedures are shown. The major types of packages or case designs used to house microelectronic circuits and how each should be mounted on the printed circuit board are examined.

The last portion of the module discusses replacement of faulty components and the mounting of conventional parts to multilayer boards.

Key technical words

- **Keying** A method of identifying a part and a board to insure proper alignment.
- **Tack-solder joint** A quick solder joint that holds the component in place until the final soldering process is completed.
- **Desoldering tip** A special soldering iron tip that heats all the device leads at one time.
- **Wave-soldered** A soldering procedure that solders all the components to the conductors at one time.

Performance objectives

Upon completion of this module, you should be able to:

- Identify the tools required in assembling a printed circuit board.
- Explain the importance of keying.
- Identify a properly mounted component.
- List the steps required to repair a faulty connection.

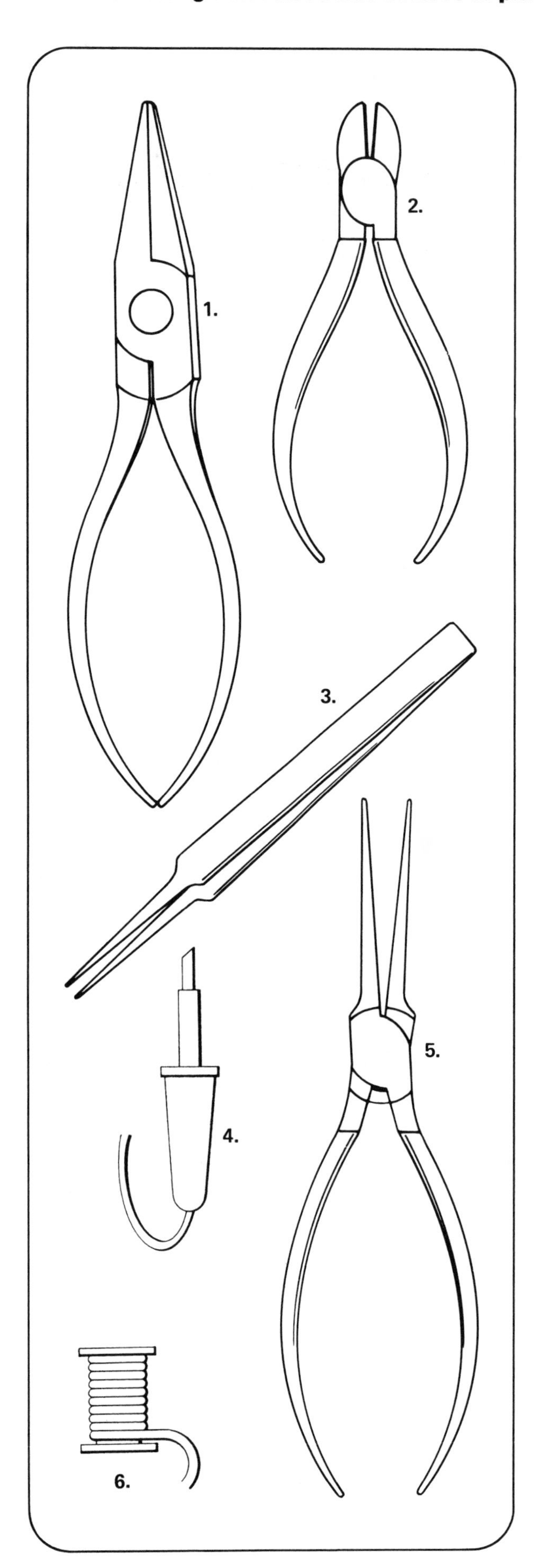

Assembling microelectronic devices to printed circuit boards

Required tools

1 Dressing pliers or chain-nose electronic pliers.

2 Side cutters.

3 Tweezers.

4 Temperature-controlled soldering iron.

5 Needle-nose pliers

6 Resin core solder (60 percent tin/40 percent lead).

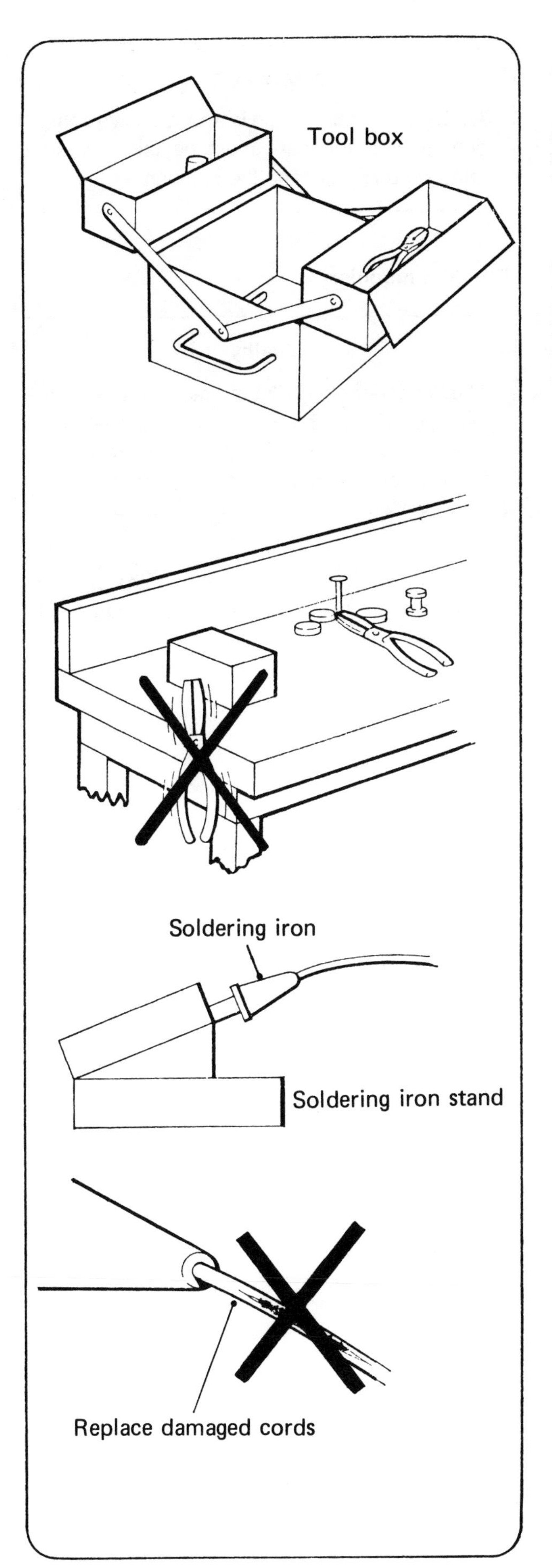

Care of tools

When tools are not in use, pack them neatly in a tool box or replace them on the tool board. Keep the tools and work area clean.

When tools are in use, place them neatly on the bench in a position where they can be easily reached and cannot be knocked to the floor.

Keep the soldering iron in a stand when it is not in use.

Safety

The soldering iron cord should be in good repair.

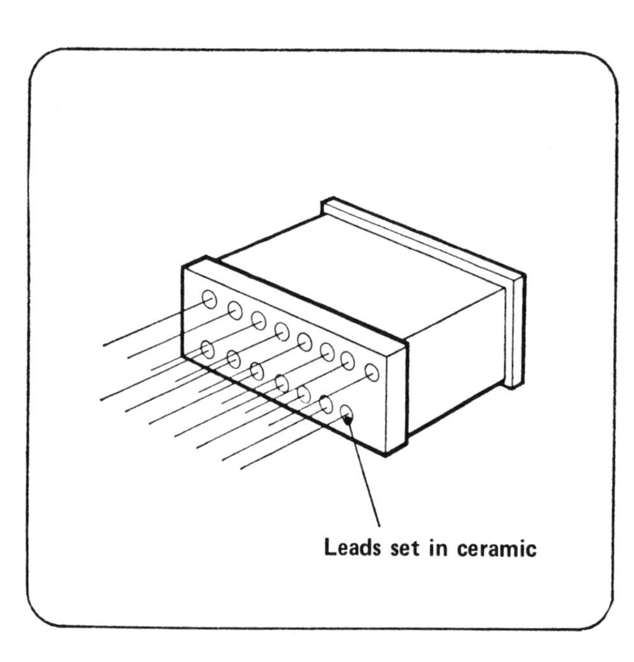

Quality

A microelectronic device is an expensive component. Extreme care must be taken when handling and mounting the component.

Thin film modules

Quality

The leads on thin film modules are set in ceramic that can be easily cracked. The pins can be easily broken. Extreme care must be taken when handling these components.

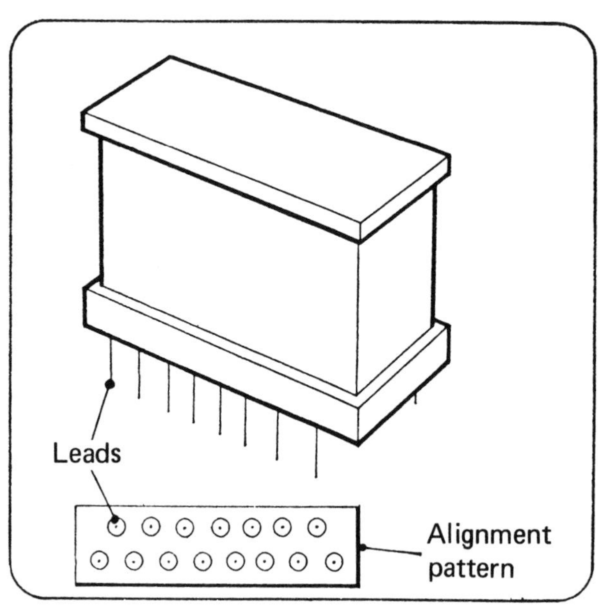

The leads are in an alignment pattern that prevents the component being incorrectly inserted in the printed circuit board.

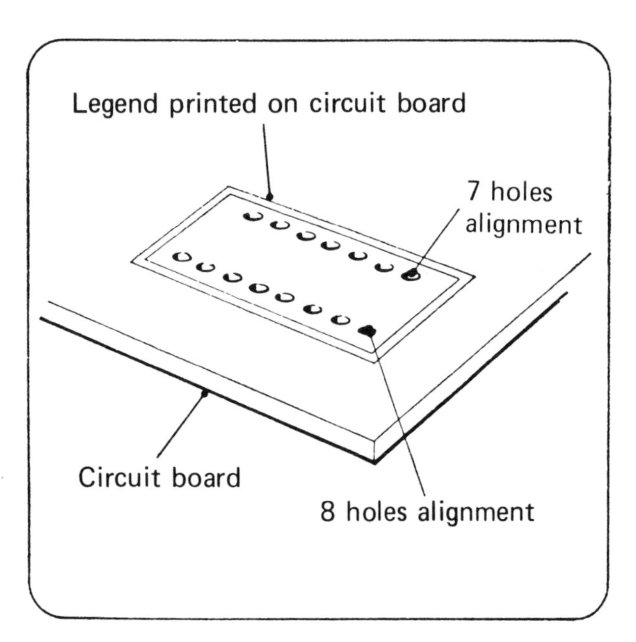

The legend printed on the circuit board corresponds to the alignment on the component's leads.

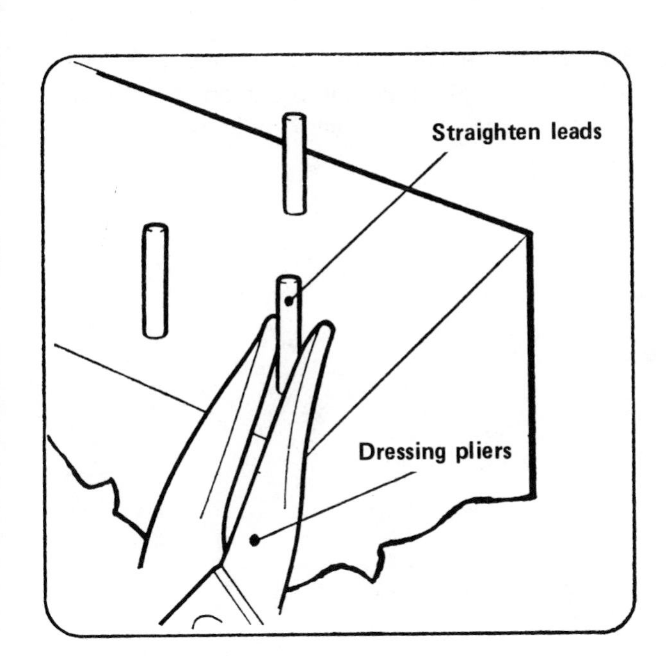

Straighten the pins, if necessary, before fitting the component to the printed circuit board. Take care not to break or bend the leads.

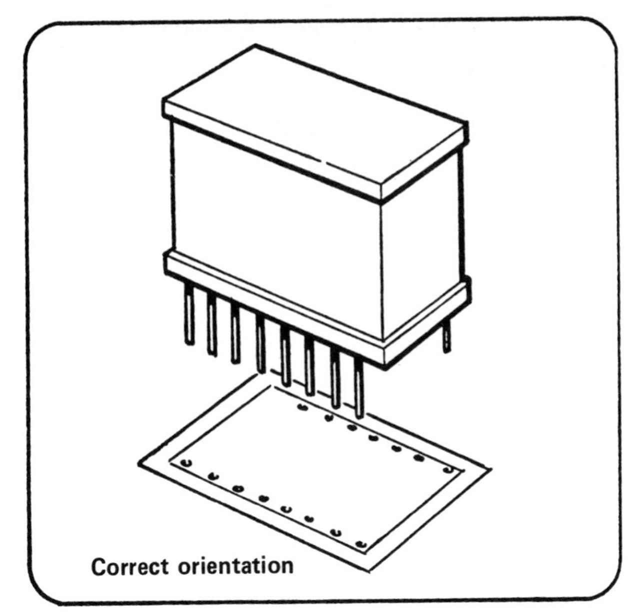

Determine the correct orientation of the components.

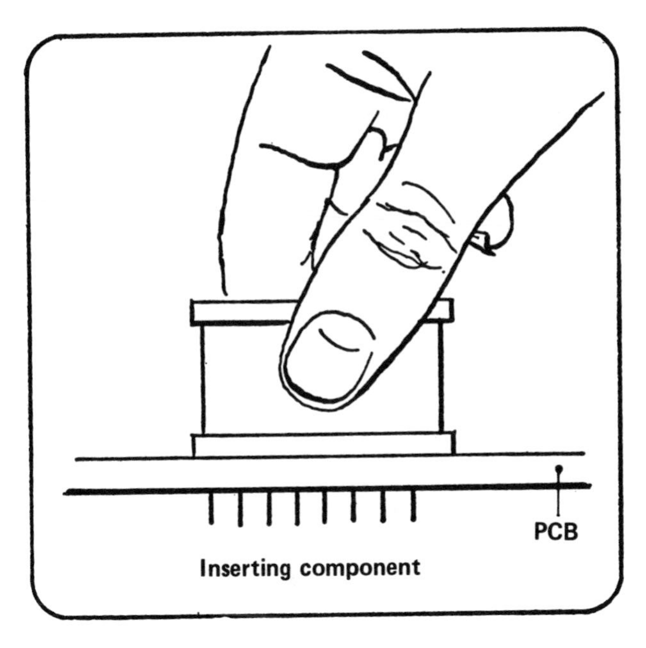

Insert the leads in the holes and push them through.

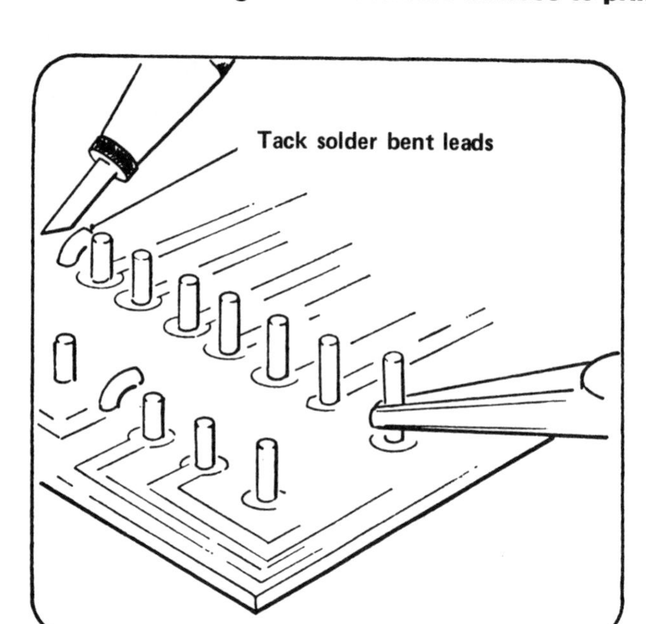

To secure the component, bend and tack-solder three leads. Take care not to damage the conductors and pads when bending the leads.

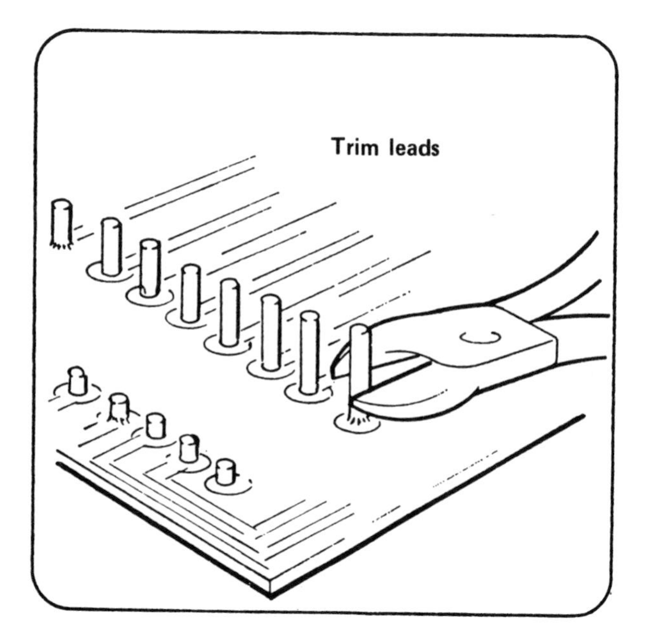

After tack-soldering, straighten the three leads.

Cut the leads off close to the board.

The component is now ready to be wave-soldered.

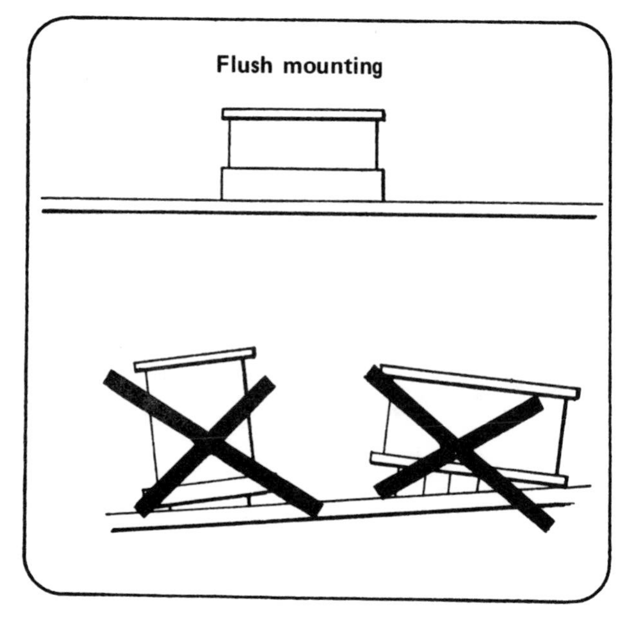

Quality

The component must be seated against the PCB.

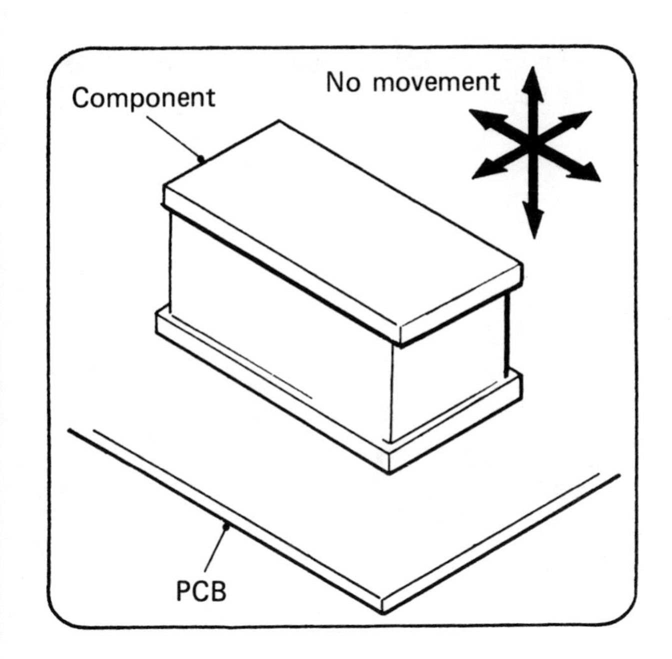

Quality

The component must be secure. No movement should be possible.

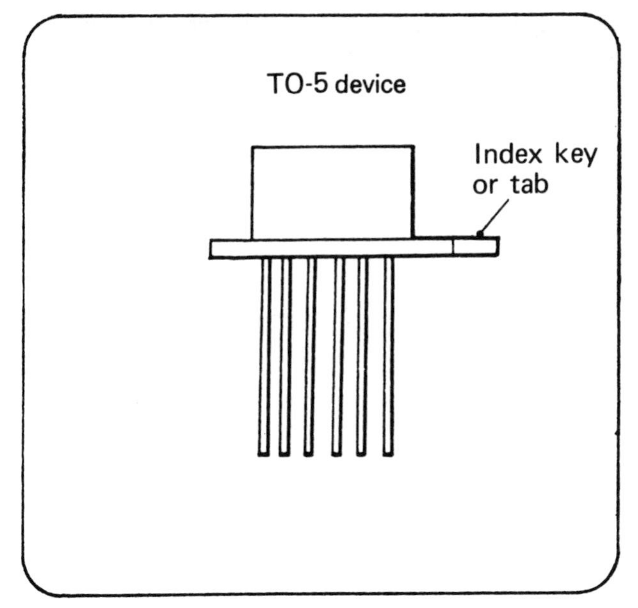

TO-5 devices

A TO-5 device is a standard mounting package for many transistors and integrated circuits. They normally have 3, 4, 6, 8, or 10 leads. Other package configurations of this type (TO-93, TO-3) will follow the same procedures.

Quality

Care must be taken when assembling these devices because of the high cost of the device and/or reworking.

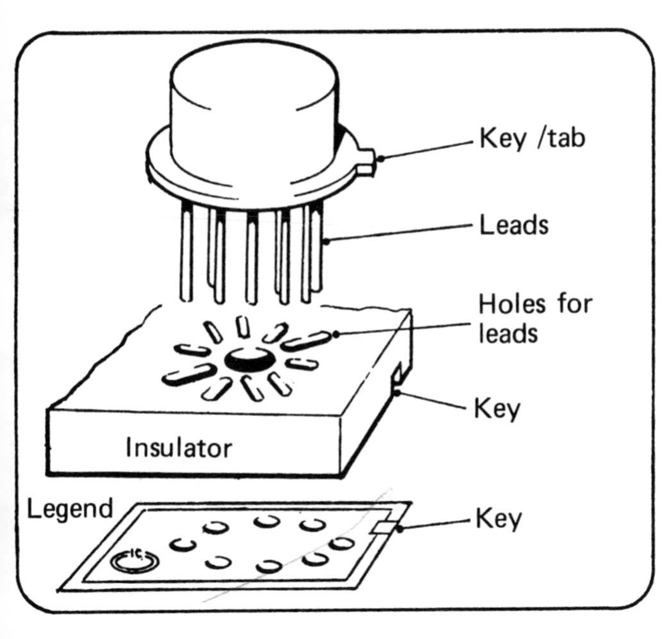

The parts involved in the assembly of the device are

TO-5 device.
Insulator/support.
Legend on printed circuit board.

Pass the device leads through the insulator so that they are just protruding through the underside of the insulator. Place the lead ends in the holes on the printed board.

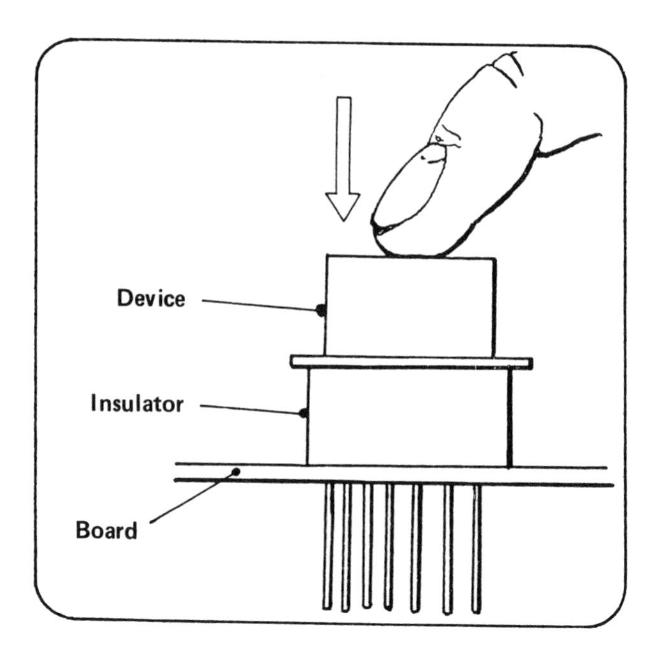

Push the device and insulator firmly down onto the printed board.

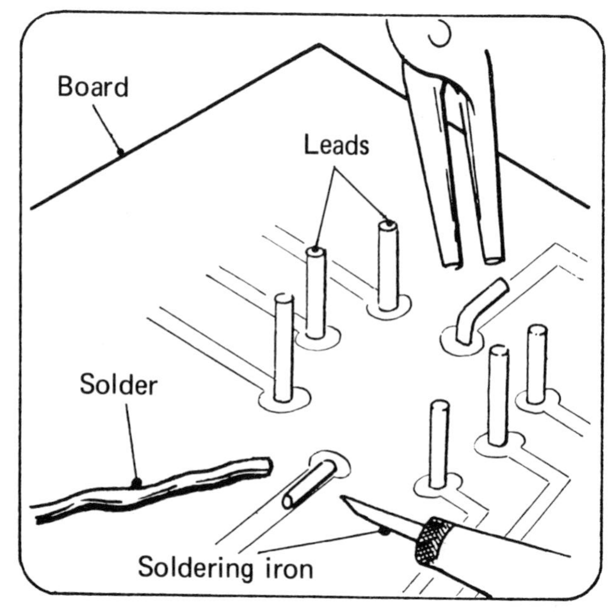

Holding the device in position with a finger, bend two leads and tack-solder them to secure the device.

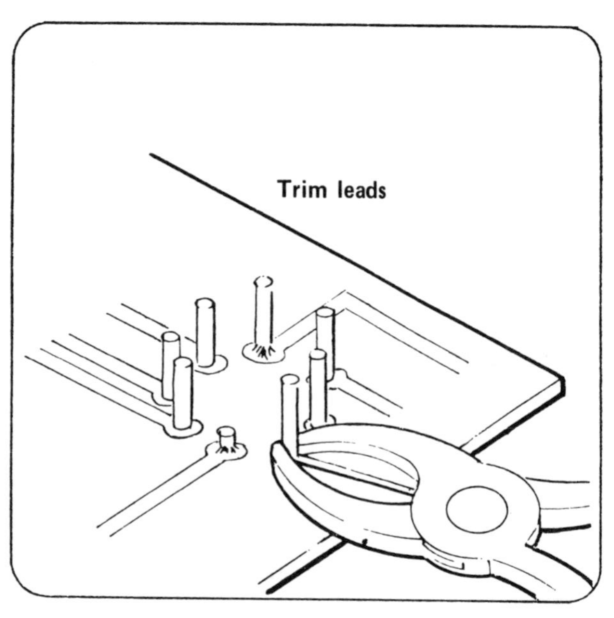

After tack-soldering, straighten these two leads and cut all leads close to the board. The device is now ready to be wave-soldered to the board.

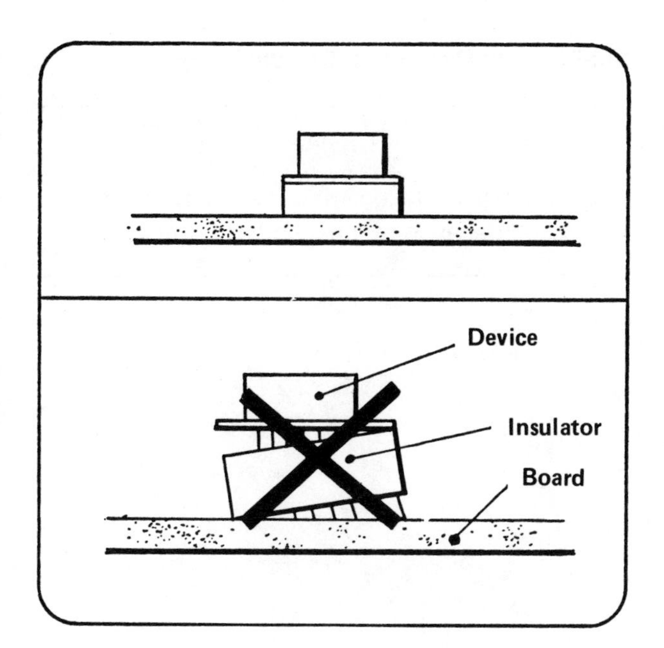

Quality

The device and insulator must be seated flat against the PCB.

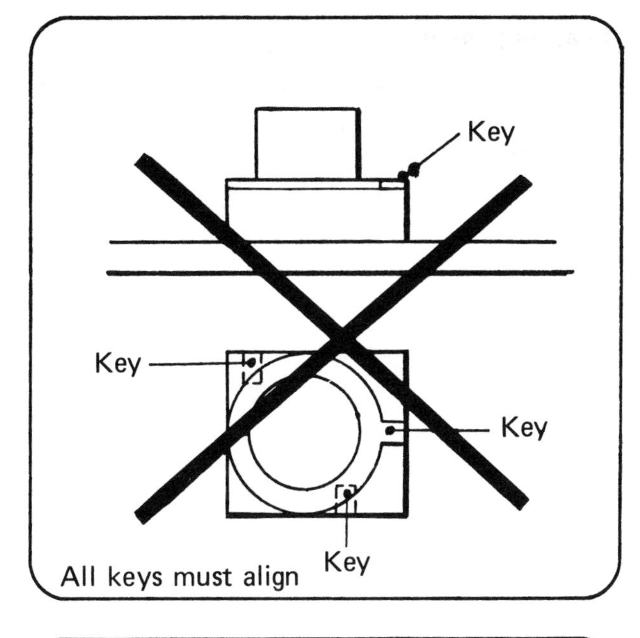

Quality

The key of the device must correspond to the key on the insulator and the key on the legend.

The alignment of the keys insures that the device is placed on the board correctly.

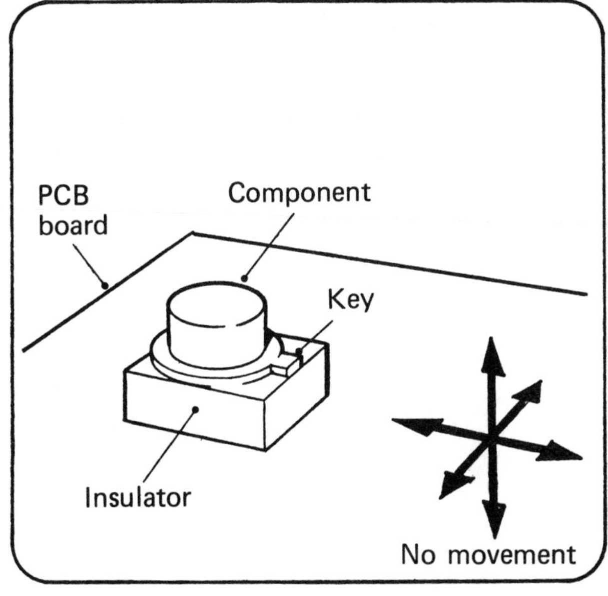

Quality

The device must be fitted tightly. There should be no movement.

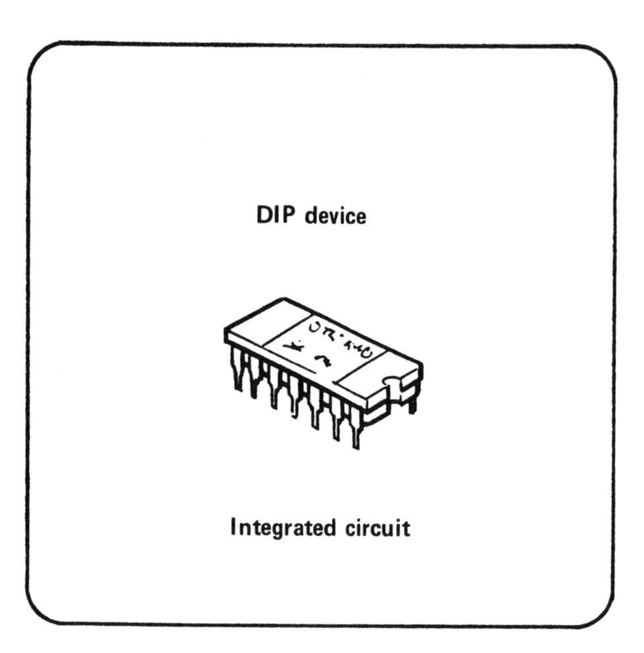

Dual-in-Line Packages (DIP devices)

> **Quality**
>
> Care must be taken when assembling these devices because of the high cost of rework.

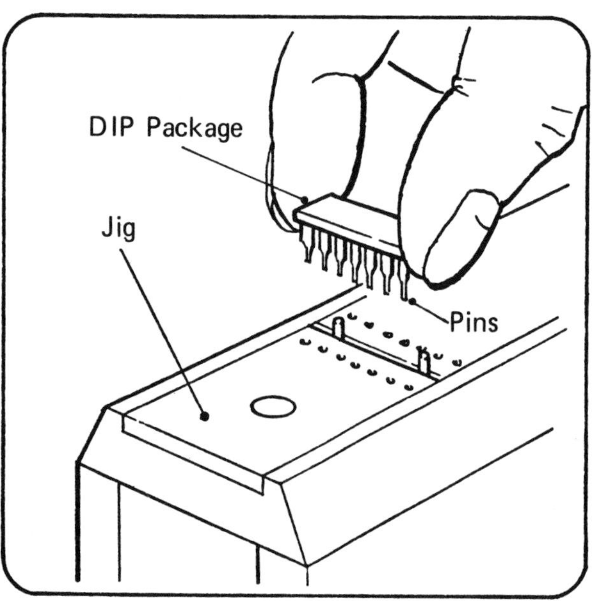

Pin arrangement

The pins must be straight before fitting the device to a printed board. This can be done by using a pin straightening jig. It is important that the pins are straight before assembly to the board.

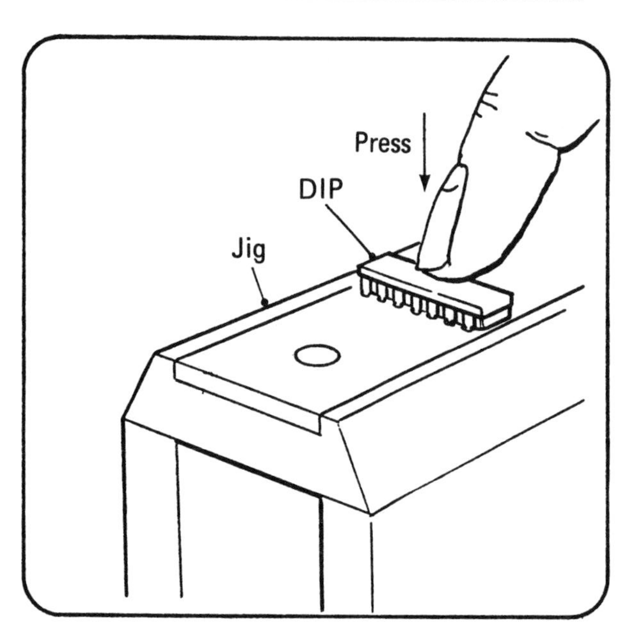

Using the pin straightening jig

1 Line the pins up over location holes in the jig. The term pin and leads are used interchangeably to indicate the conductive material from the device to the board.

2 Accurately align the pins with the location holes and push down firmly. A spring release will eject the device.

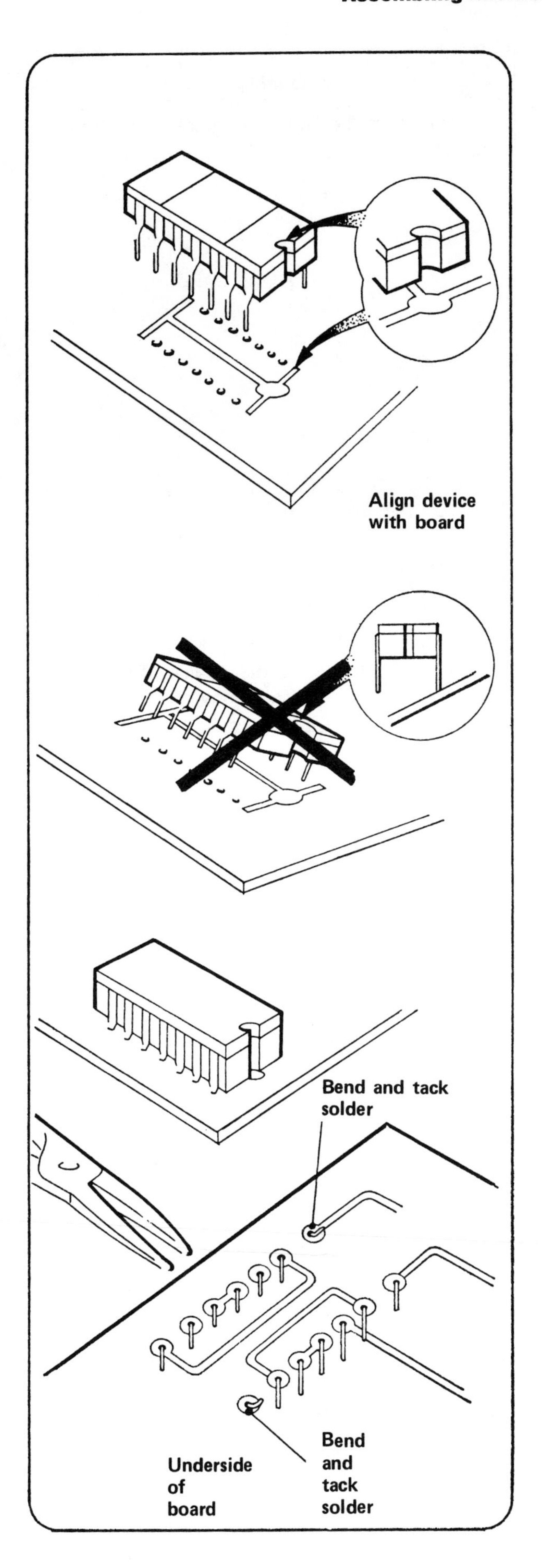

Assembly to printed board

1 Line the device pins up with the holes in the printed board. Be sure that the device polarity (GROOVE) and the legend polarity correspond.

2 Fit the device carefully so that the pins insert fully.

3 The device should fit flat against the board.

4 Secure the device by bending two leads outwards at a 45° angle as specified. Tack-solder bent leads and cut all leads close to the board. The device is now ready for wave-soldering.

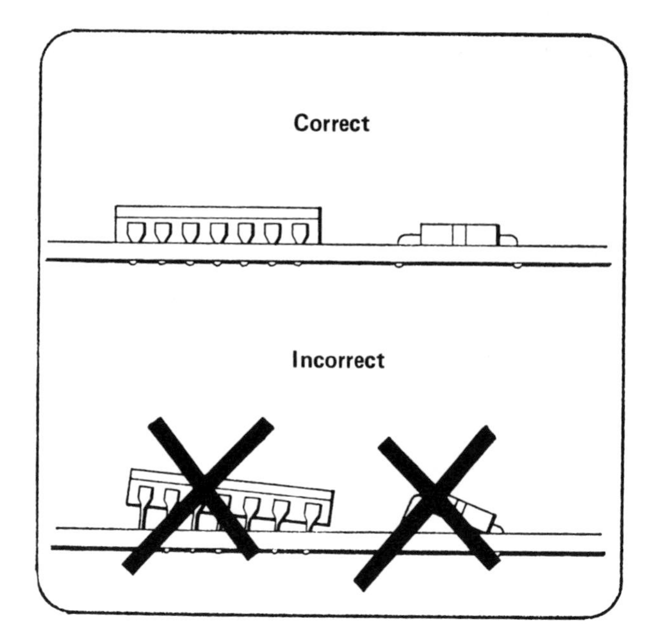

Quality

Devices must be fitted flat against the board.

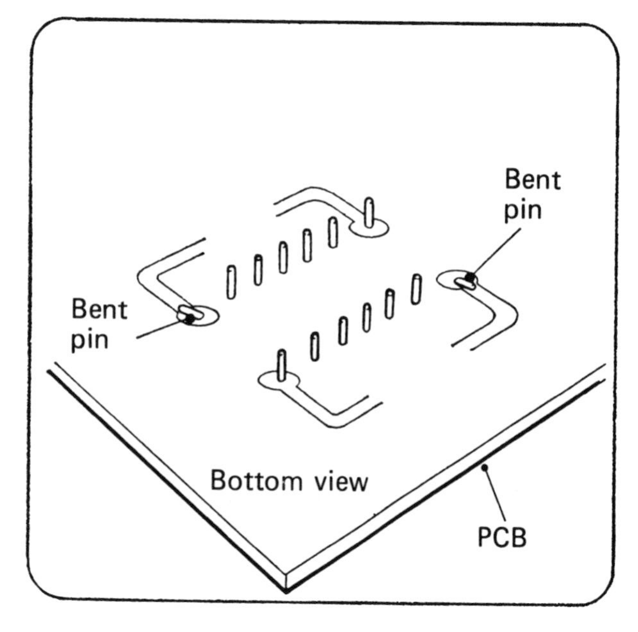

Quality

Only the number of pins specified need be bent at 45° to secure the device. Two pins have been bent in this example.

Quality

The device polarity and legend polarity must correspond. The polarity generally indicates the first pin.

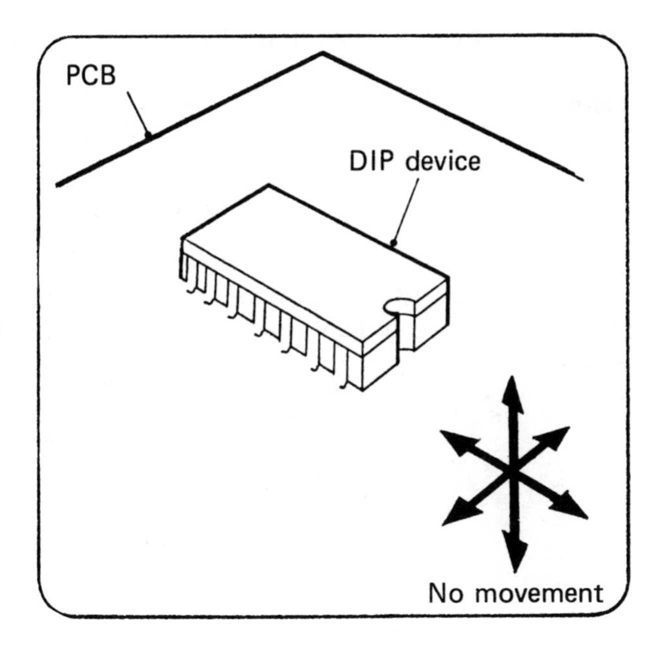

Quality

The device must be fitted tightly. There should be no movement.

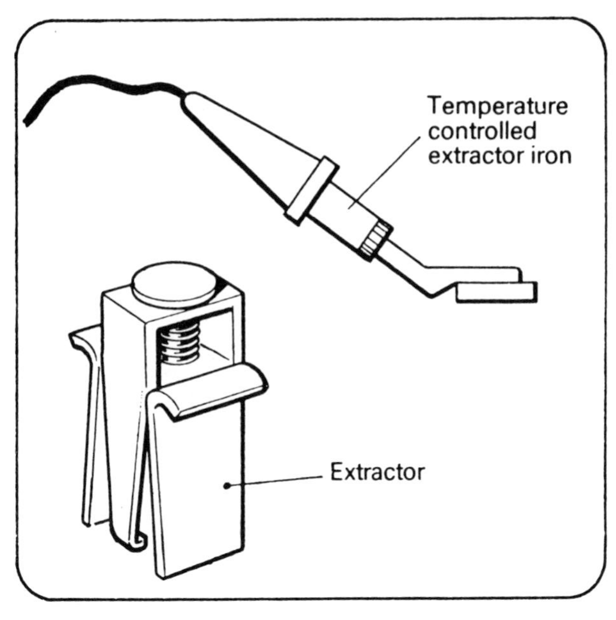

Removing DIPs from PCBs

If a device is found to be malfunctioning when the board is tested, it has to be removed and replaced.

Extraction tools

1 Extraction iron or desoldering tip.

2 Extractor.

Quality

Care must be taken not to damage the PCB when extracting a device.

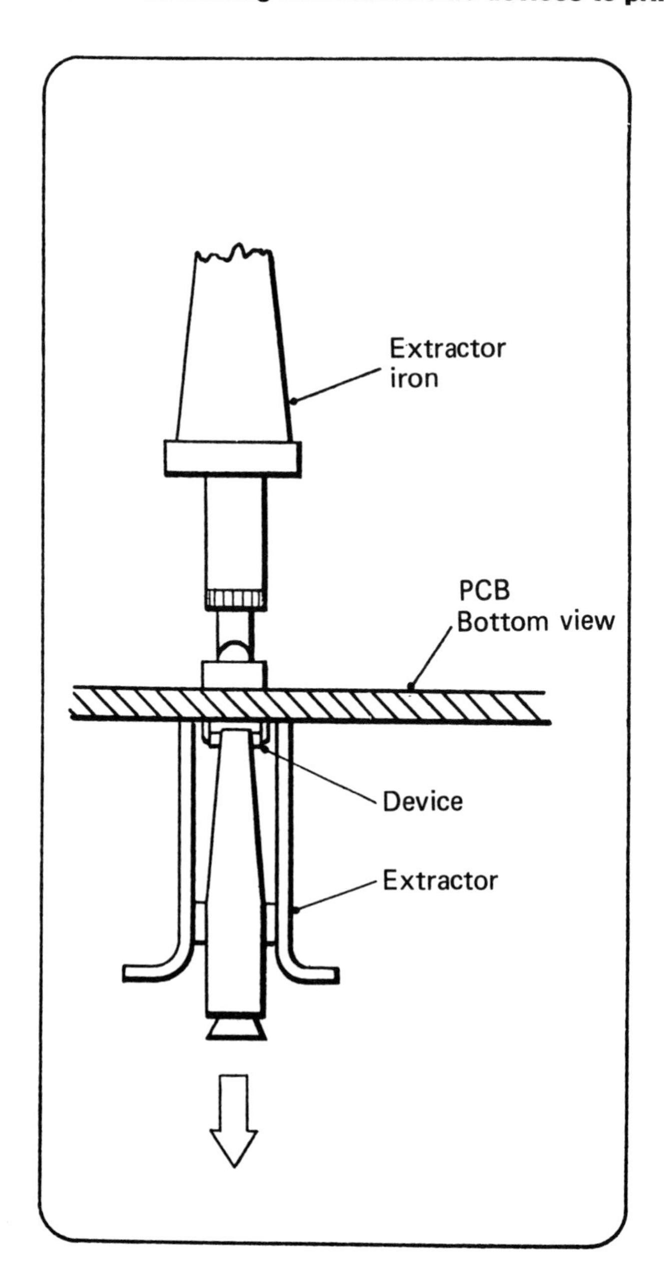

Methods of extracting a DIP device

Method 1

1 Clamp the extractor to the device.

2 Place the extraction iron over the soldered pins of the device.

3 When the solder softens the extractor will withdraw the device and drop away taking the device with it.

4 Remove the solder left behind.

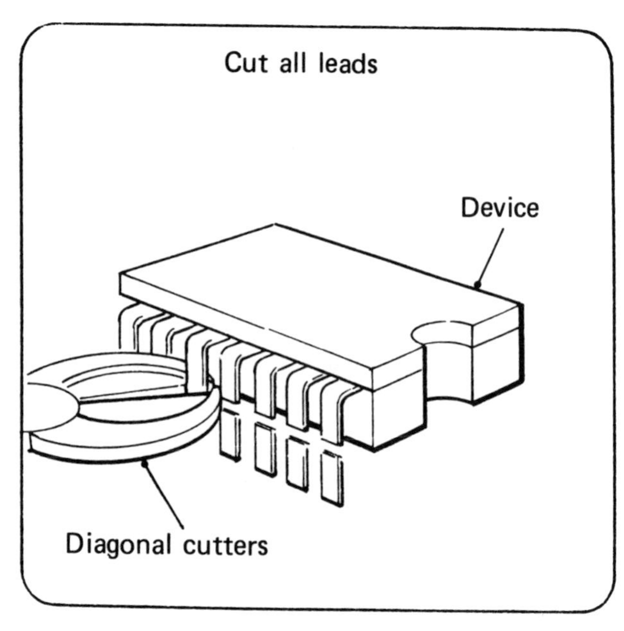

Method 2

1 Cut all leads on component side of the board.

2 Reheat each joint and withdraw each lead.

3 Remove excess solder and insure holes are clear.

> **Quality**
>
> Take care. Prolonged heat applied to the conductor may cause damage.

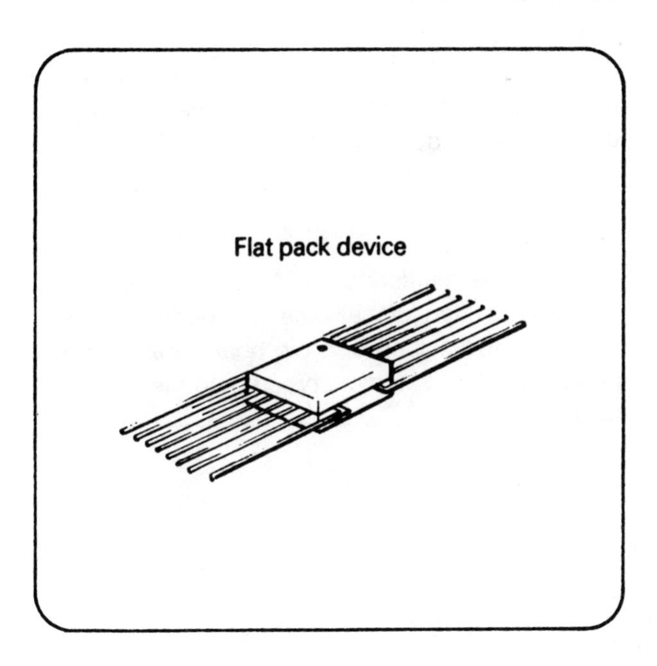

Flat pack devices

> **Quality**
>
> Care must be taken when handling the devices because of their high cost.

Flat pack devices can be assembled in the conventional method as DIP devices, that is, the pins go through the board and are wave-soldered.

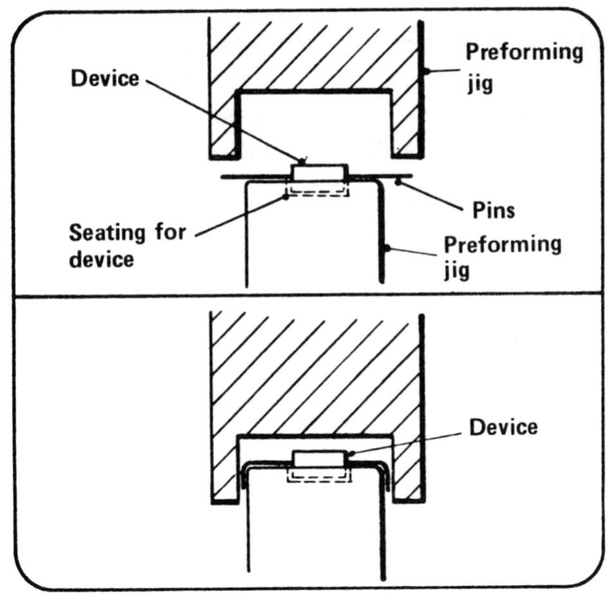

The technician preforms the pins using a jig.

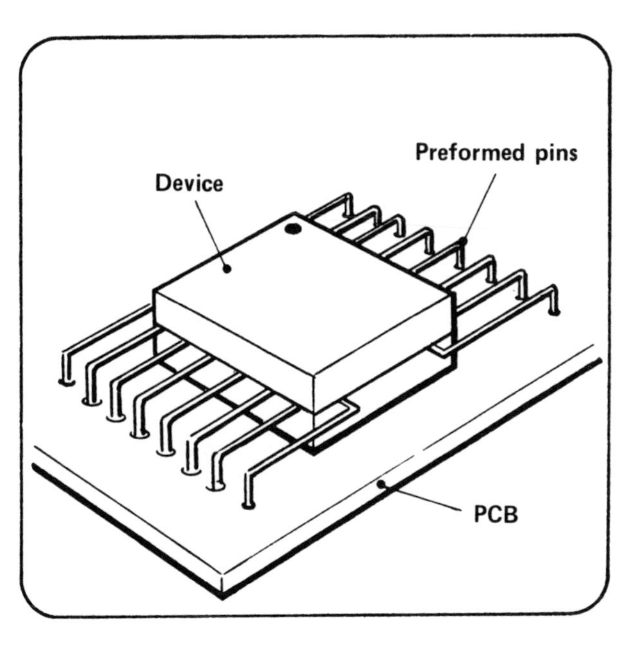

The device is fitted to the board, tack-soldered, and then wave-soldered.

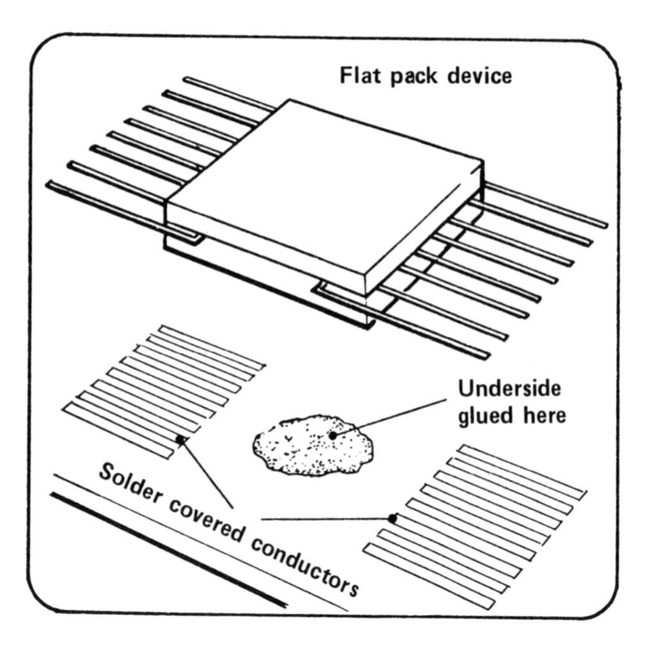

Another method is shown. The device is glued in position on the printed board. The device leads must overlay the conductors exactly.

Quality

When this method is used, the flat pack devices are fitted after all other components have been wave-soldered to the board. Insure the surplus glue does not spread onto the track.

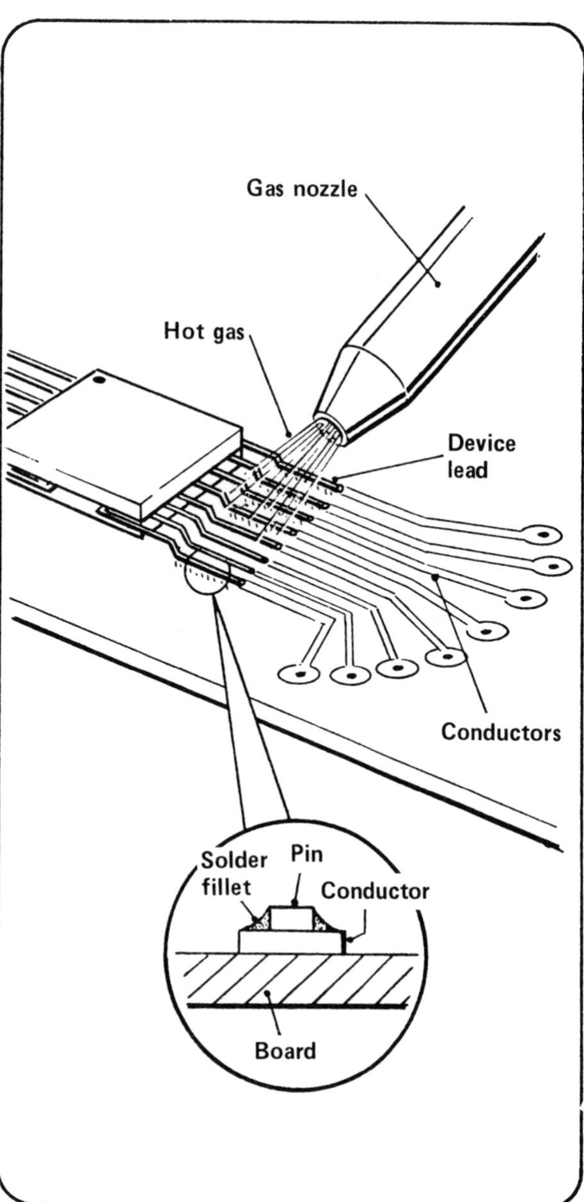

The pins are soldered to the conductor by a re-flow method. In the example shown, hot gas is being used. Other re-flow methods used to form the joint for this method of fitting flat pack devices include

1 Infrared rays.

2 Hot oil—the complete board is immersed in hot oil.

3 Electrical pulses.

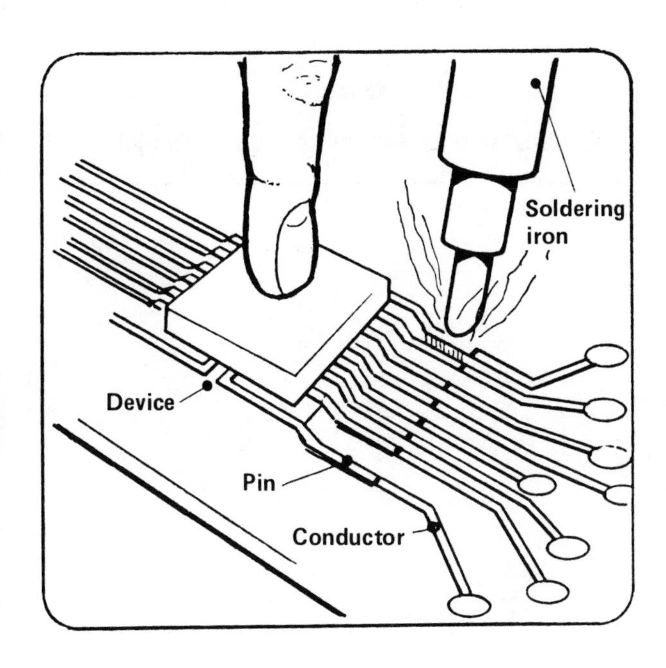

Hand-soldering method

Flat pack devices can be hand-soldered to the board.

1 Place the device in position and hold.

2 Secure the position of the device by running the soldering iron along two or three of the device's pins.

3 Solder the remaining leads to their conductors.

4 It may be necessary to insert a heat sink between the pin being soldered and the device.

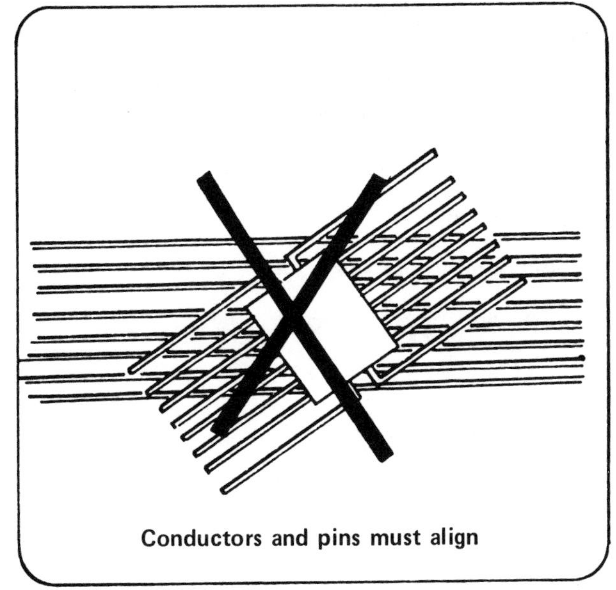

Conductors and pins must align

Quality

Soldered conductors and device pins must align properly.

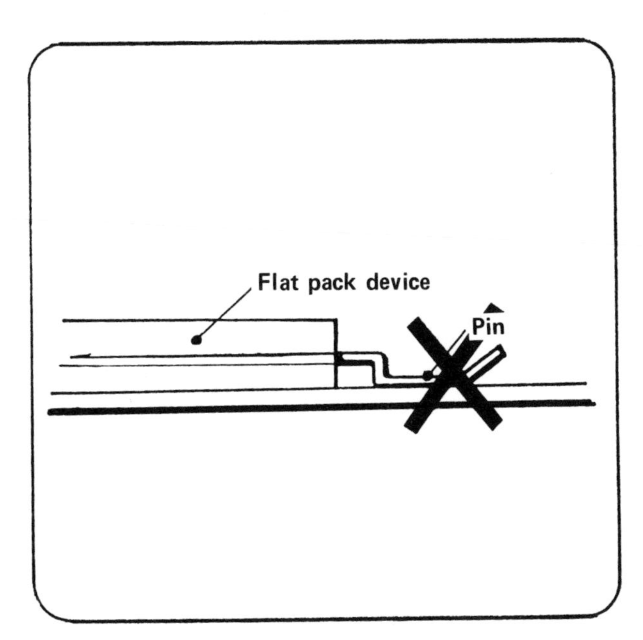

Pins must not be lifted. They must be fully attached to the conductor.

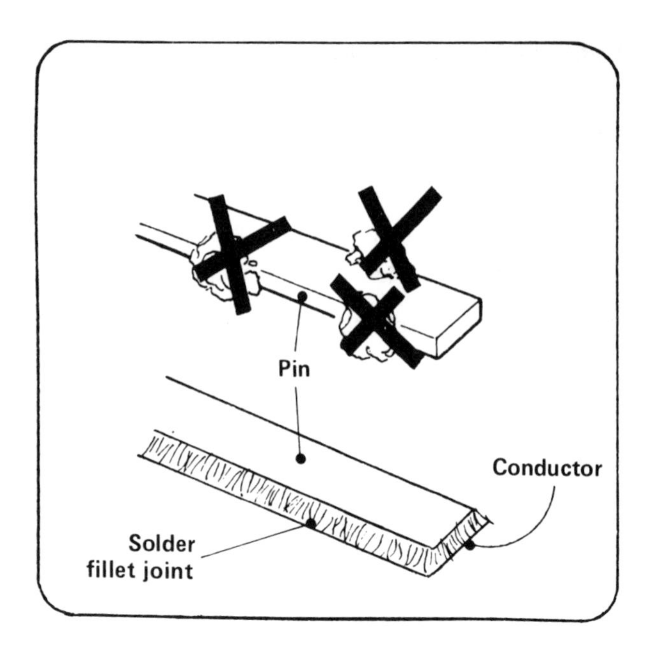

Quality

The soldered joint must be a good fillet joint.

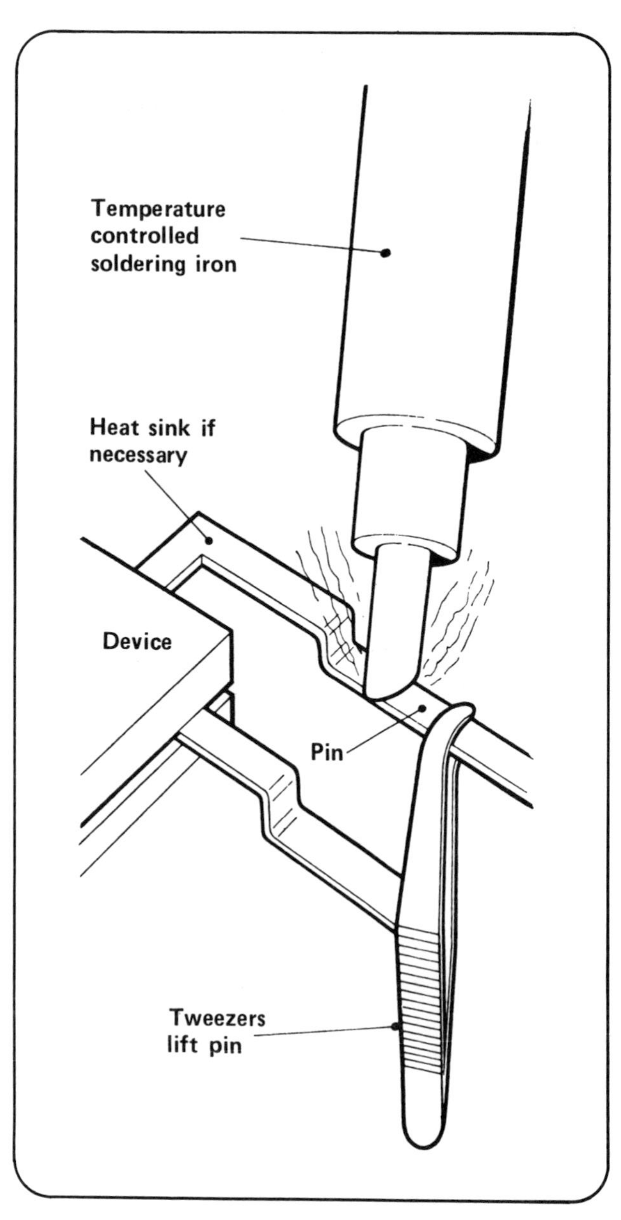

Repairing a faulty connection

If a fault occurs owing to a substandard soldered joint, the joint is easily repaired.

1 Heat the faulty joint.

2 Lift the pins.

3 Remove the old solder.

4 Resolder the conductor.

5 Hand-solder a new joint.

Quality

Care must be taken not to damage the conductor or board.

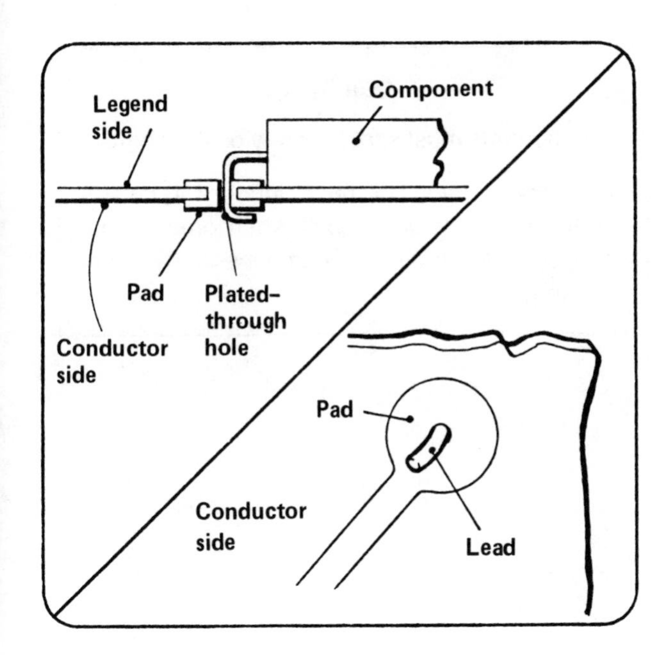

Assembling conventional components to through-plated and multilayer boards

Components are fitted to plated-through and multilayer boards in a similar manner to those on a normal printed board. The component lead is fed through the board from the legend or component side. The lead is bent in line with the conductor. The lead is cut to fit the pad.

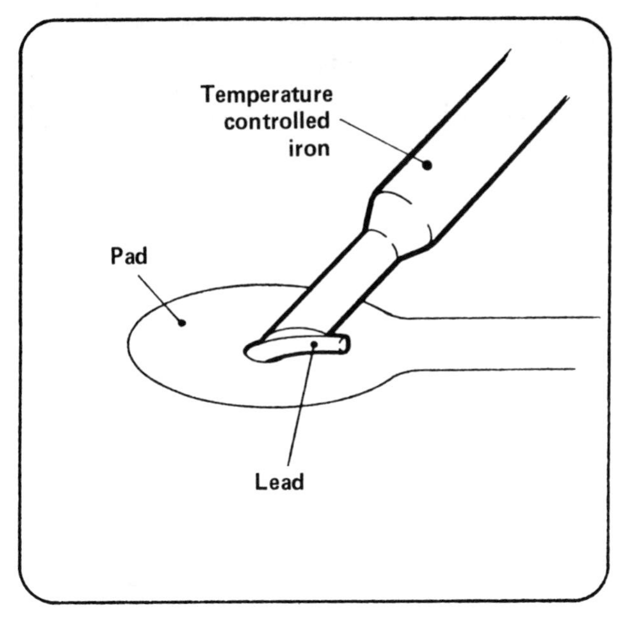

Hand-soldering

Always use a temperature-controlled soldering iron at the specified temperature. Solder the joint as quickly as possible.

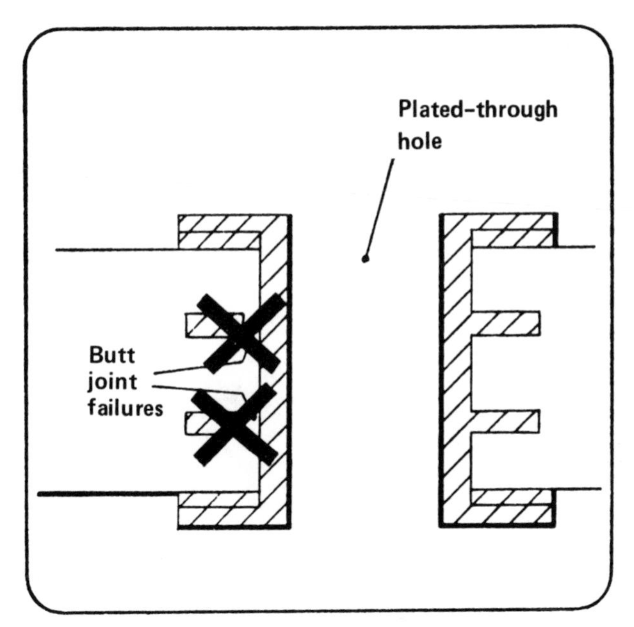

Quality

The glass/epoxy resin and the copper conductors on a multilayer board expand at different rates when heated.

Butt joint failure may occur at the inner pads if the board is overheated by the soldering iron.

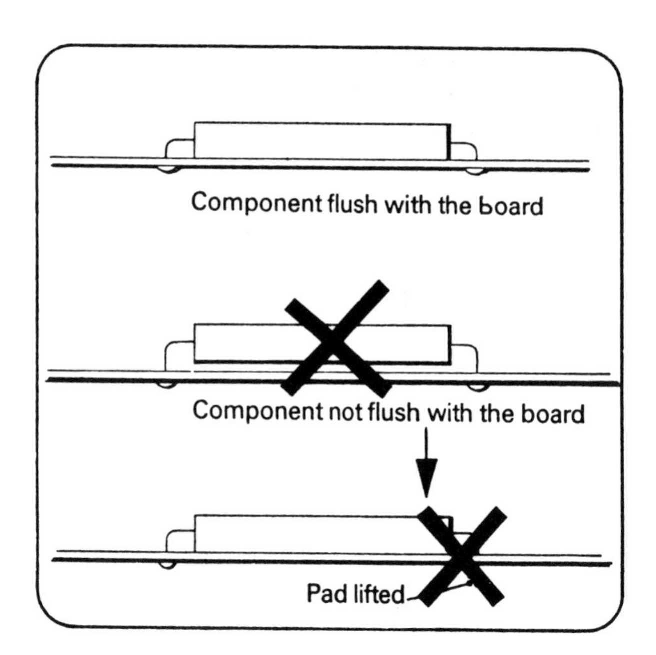

Quality

Components must seat squarely on the board.

Incorrect seating of component will cause strain on the soldered joint when pressure is applied to the top of the component. This can cause pads to lift.

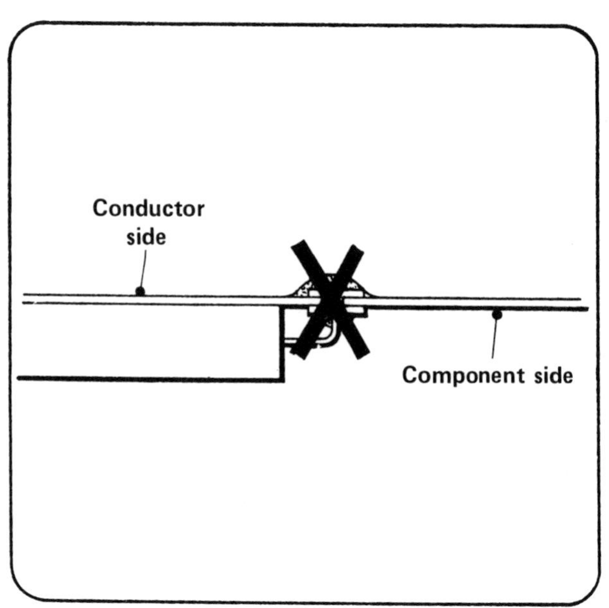

Quality

There must be no buildup of excess solder on the component side of the board.

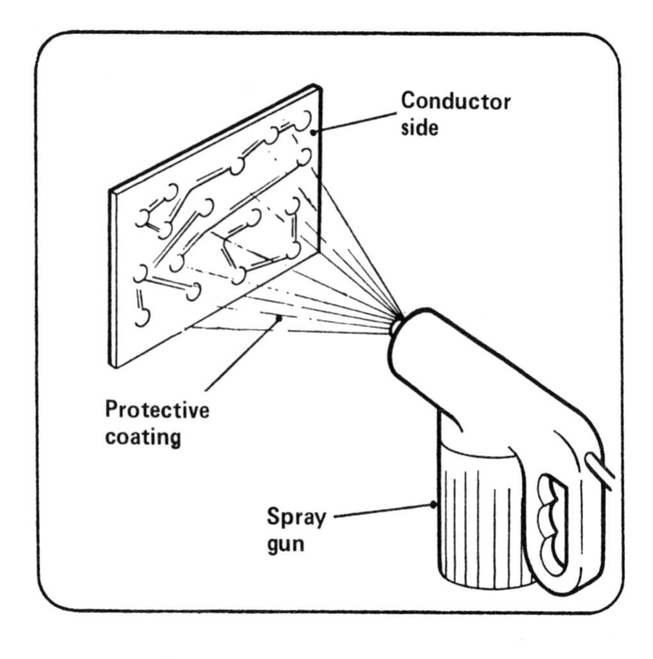

Protective coating

After assembly, some boards are sprayed on the conductor side with a protective coating. This is usually an epoxy resin or varnish depending on the customer's requirements.

Test items

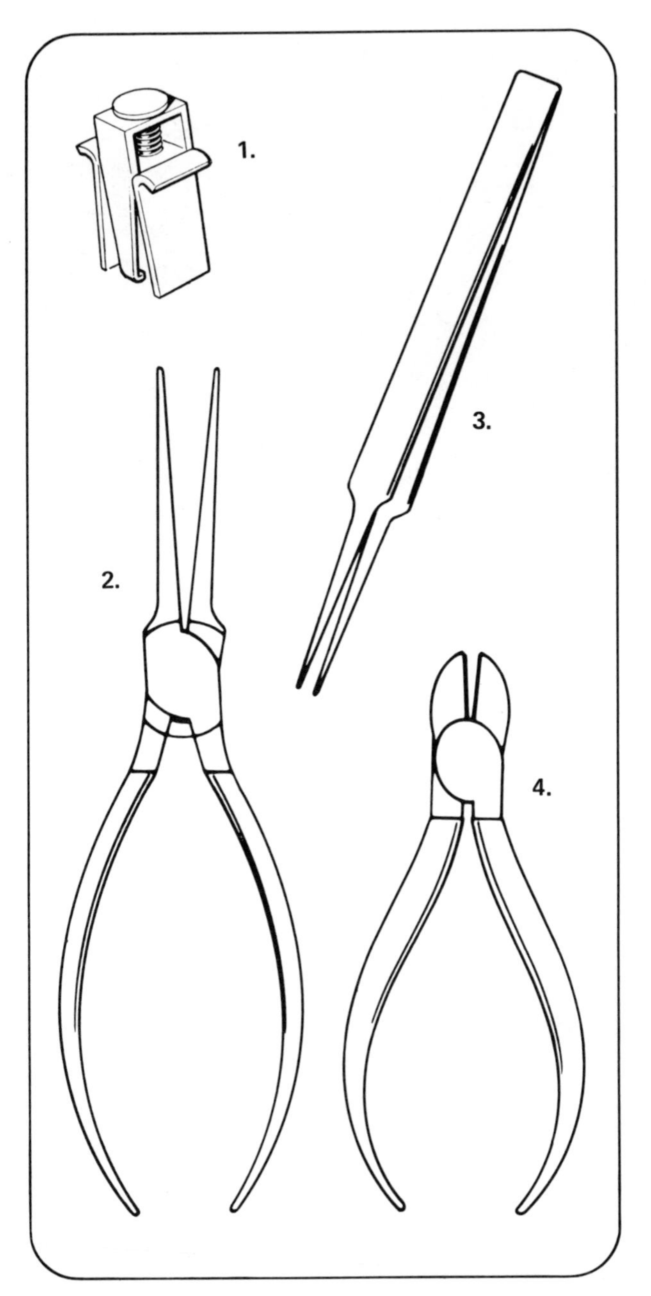

1 Match the tool names on the right with the correct drawing.

1	______	A	Tweezers
2	______	B	Needle-nose pliers
3	______	C	Side cutters
4	______	D	Extractor

(See page 88.)

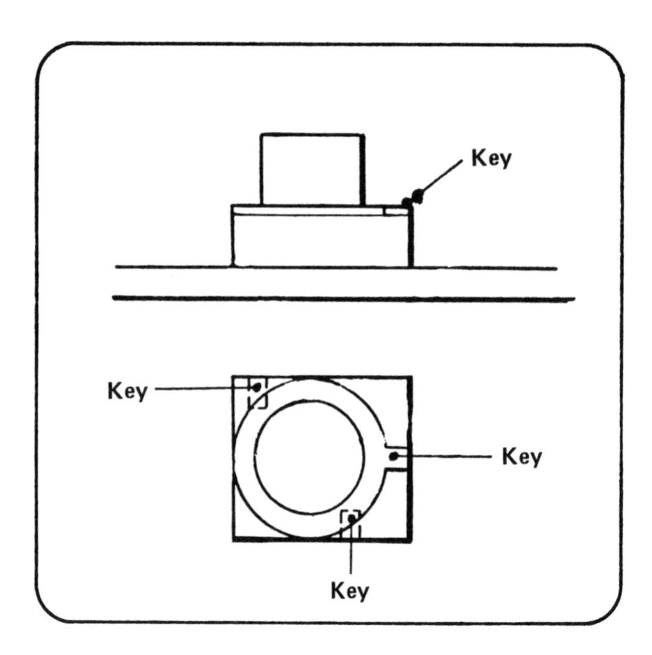

2 Using the illustration, explain keying. Indicate on the illustration how parts should fit.

(See page 95.)

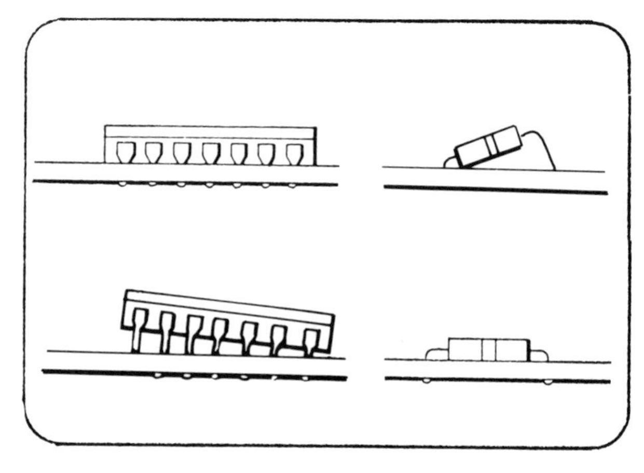

3 Given the illustration, circle the properly mounted components.

(See page 98.)

4 Complete the following list of the steps required to repair a faulty connection on a PCB

(a) Heat the faulty joint.

(b) _______ the pins.

(c) _______ the old solder.

(d) Resolder the conductor.

(e) Hand-solder a new joint.

(See page 104.)

5 Tack-soldering is used to

(a) Make a permanent solder joint for all the device pins.

(b) Make a temporary connection designed to hold the component in place until the final soldering process.

(c) Bend the leads.

(d) Correct a faulty joint.
(Circle one.)

(See page 98.)

Module nine
Subassemblies

Introduction

This module examines the mounting of subassembly printed circuit boards to the mother board. The subassemblies make use of color coding techniques, vertical and horizontal mounting, as well as flexible conductors. Subassemblies expand the use of printed circuit boards and allow fast trouble shooting and repair of the main system.

Key technical words

- **Subassembly** A small printed circuit board that connects to a mother board to complete the system.
- **Mother board** The main board that usually supplies power, ground, and other common connections to subassemblies.

Performance objectives

Upon completion of this module, you should be able to:

- Explain the relationship of the subassembly daughter board to the mother board.
- List the advantage of color coding subassembly boards.
- List the ways that subassembly boards may be mounted to the mother board.

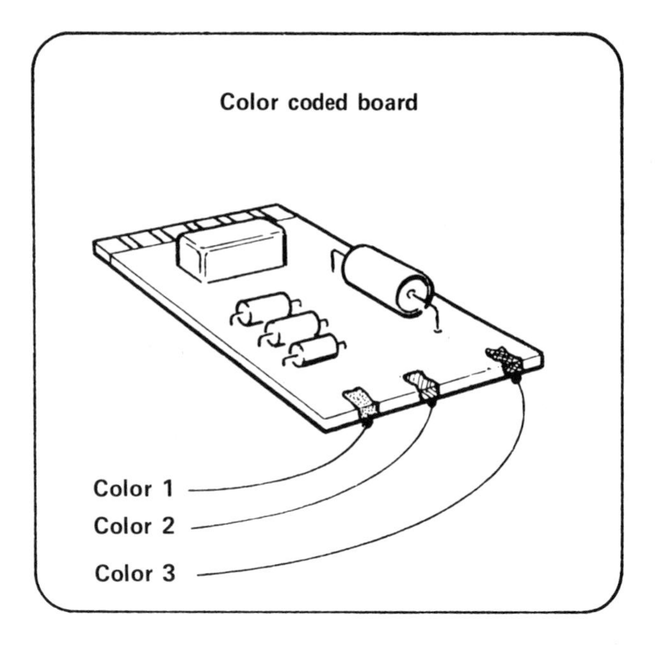

Subassemblies

Printed circuit boards

For rapid identification of circuit boards and subassemblies, a color code is often used in addition to the part number. Some boards have handles attached and the handles are coded.

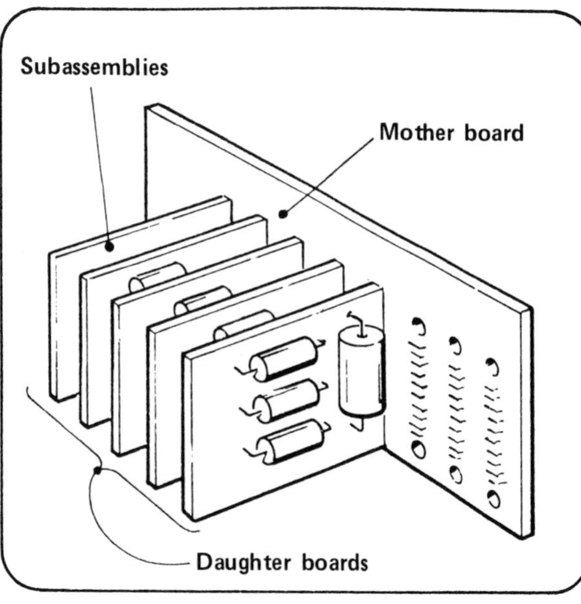

Composite PCB assemblies

Composite assemblies of PCB's are made up by mechanically fixing together two similar boards to form one assembly, by mounting smaller PCBs on a larger board, and by mounting a large number of PCBs onto a mother board. The only function of the mother is to provide connections between the daughter boards or modules and the composite assembly and external circuits. This arrangement may be known as a "nest" or a "rack" of PCBs, but many other terms are also used. Make a practice of using the local terminology but be aware that others may not fully understand your particular term.

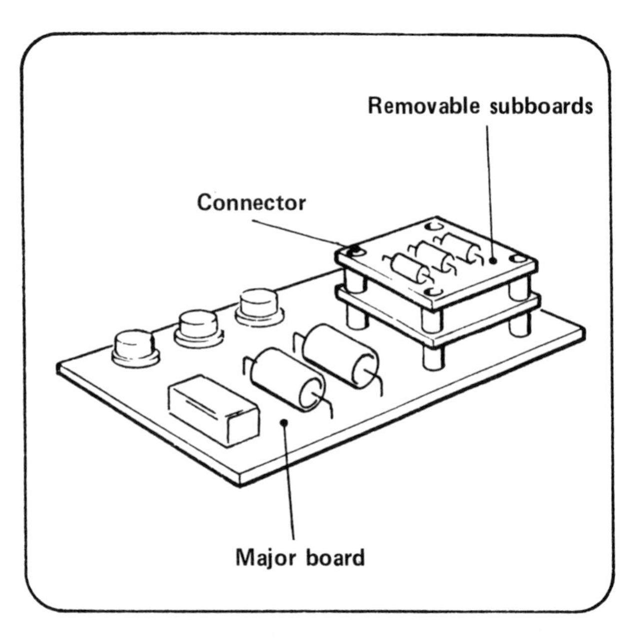

Subboards can be mounted on a major board to lie parallel with the main board. Electrical connections are made by plug and socket connectors that are arranged to allow subboards to be stacked one above the other. The subboards are removable.

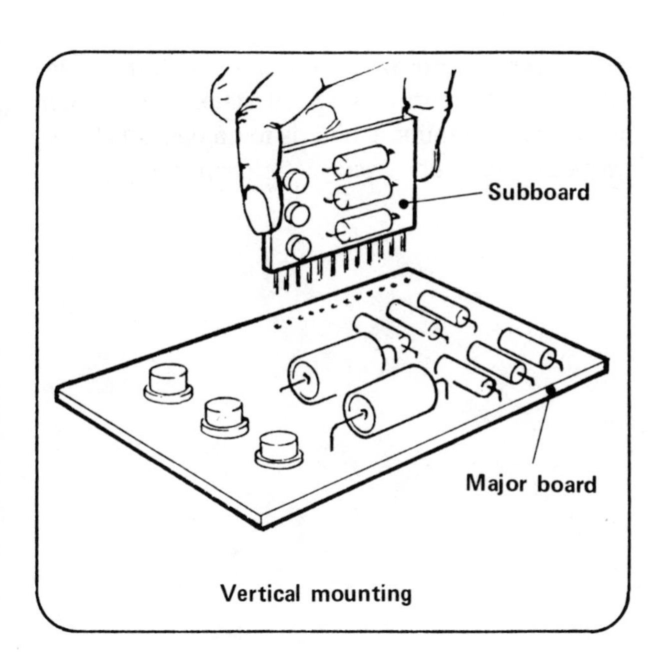

Vertical mounting

The same type of connectors can be used to mount a subboard vertically onto a master board. In this case, the subboards may be mounted side-by-side.

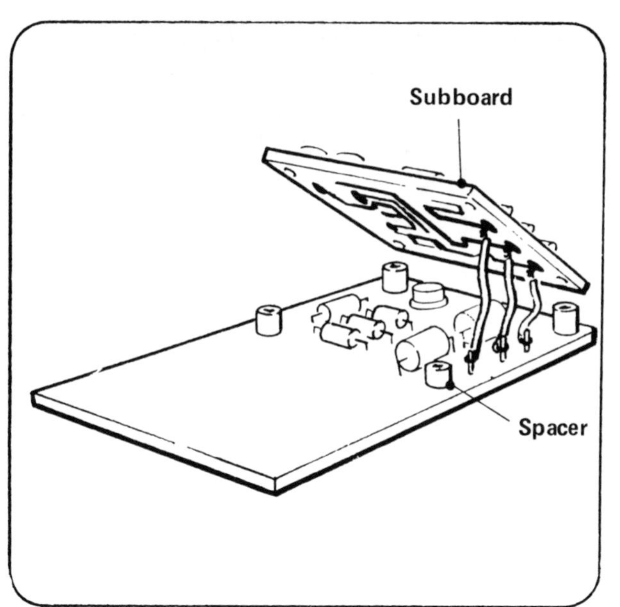

Subboards may be permanently attached to the major board by screws and nuts, with spacers maintaining the required distance between the boards. This type of mounting allows for direct soldered wire connections between the boards.

When mounting this type of subboard, the technician must take care that the screws are tightened evenly and not overtightened. Too much force could cause the boards to crack.

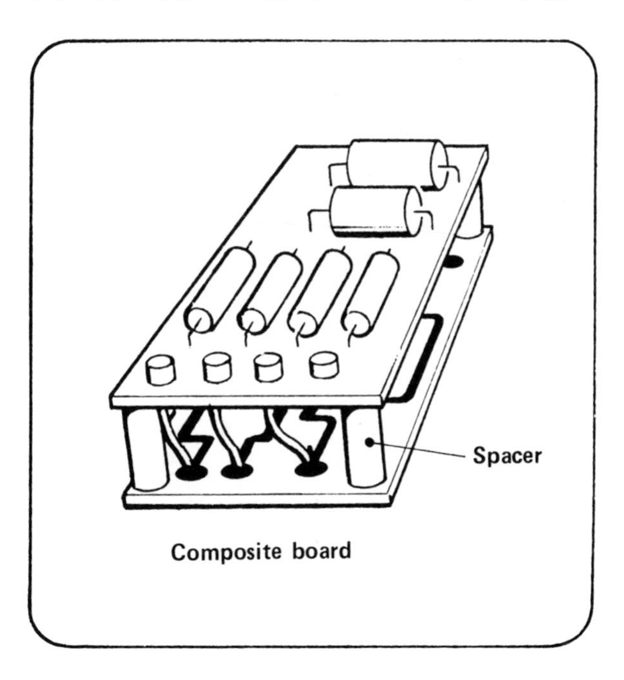

Composite board

Two boards of the same or similar size can be bolted together with the correct spacers to form one composite assembly. Interconnections can be made directly between the two.

The two end connectors are mated with a double socket connector through which connections to external circuits are made.

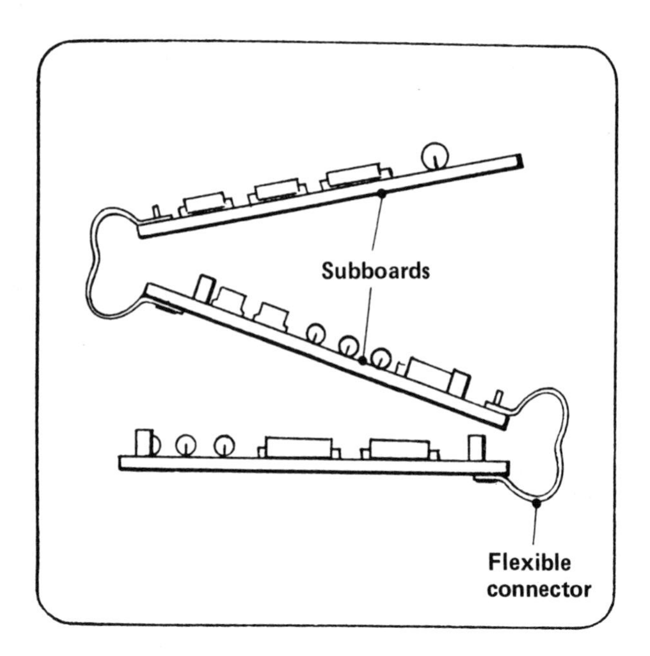

A number of similar boards can be mounted together in the form of a folded stack. This system allows for a number of boards to be packed closely while providing easy access for servicing.

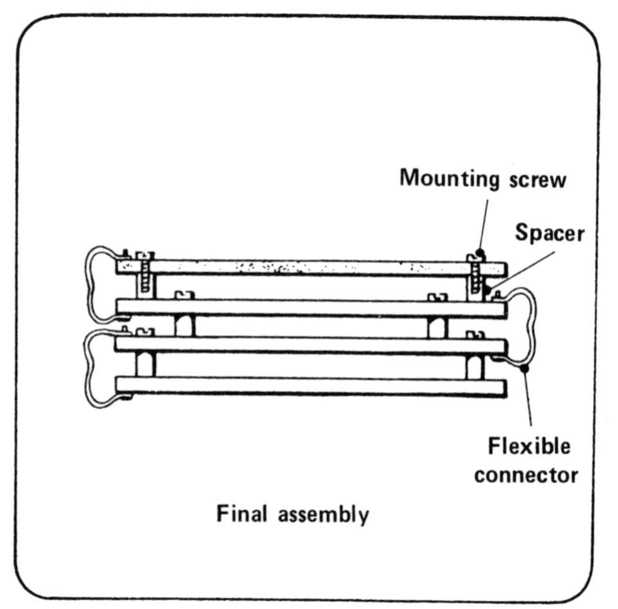

Final assembly

Flexible connectors join the conductors of one board to the conductors of both adjacent boards. These flexible connectors must allow the boards to be physically separated while still remaining electrically connected. When stacked, each board is fixed to the one below it by screws.

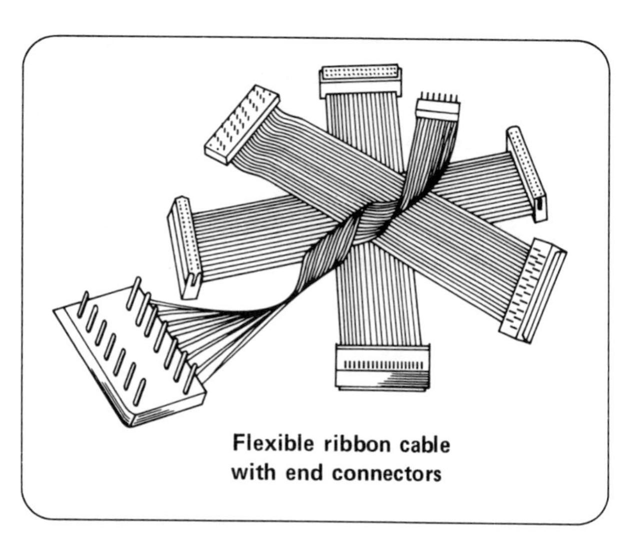
Flexible ribbon cable with end connectors

Connections can also be made between subassemblies by using a flat bonded or ribbon cable. The ribbon cable may be attached to the boards by various types of connectors. The ribbon cable is color coded.

Test items

1 List three methods of mounting subassemblies.

(a) ______________________________

(b) ______________________________

(c) ______________________________

(See page 110.)

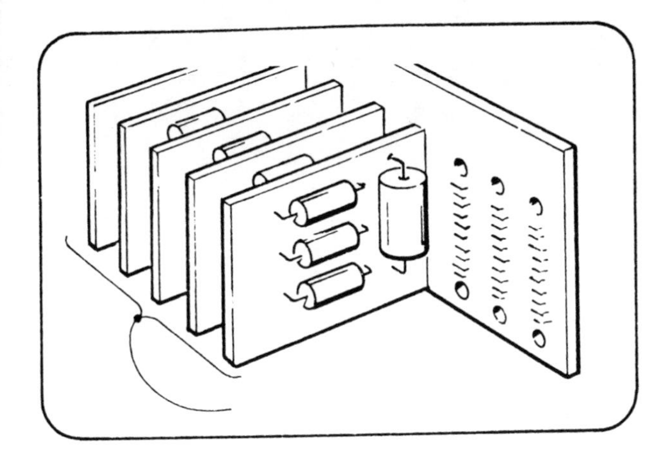

2 List the advantage of using a flexible conductor to form a folded stack of subassemblies.

____

(See page 112.)

3 Given the illustration, identify the mother board and the daughter boards.

____

(See page 110.)

Module ten
Printed circuit board connectors

Introduction

This module examines three commonly used methods for making connections to the printed circuit board. While only three methods are shown, there are many variations of edge connectors available.

Key technical words

- **Edge connector** A connecting device that allows the conductors of a printed circuit board to be easily connected with another assembly.

Performance objectives

Upon completion of this module, you should be able to:

- Identify a wire pin connector, an edge connector, and a tag connector.

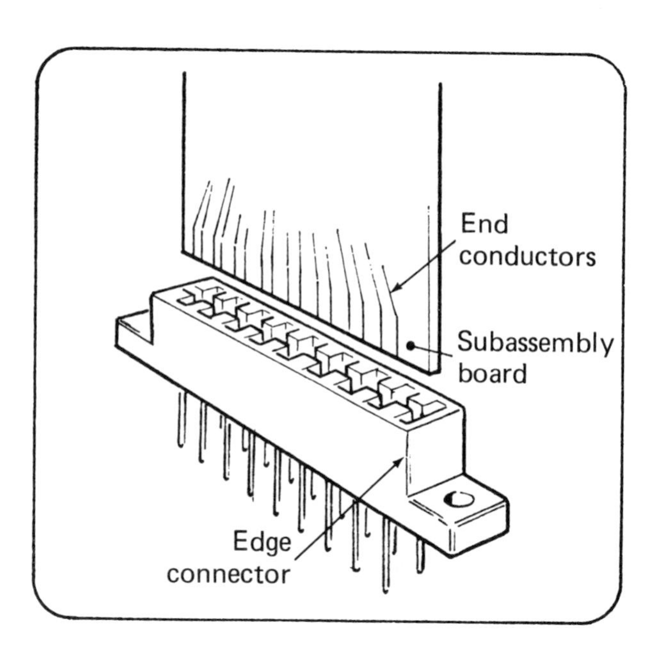

Edge connectors

Board end connectors are an integral part of the printed circuit with the conductor continued to the edge of the board. On some boards, the conductors are wrapped around the edge of the board. Double-sided boards will have the end conductors on both sides of the board.

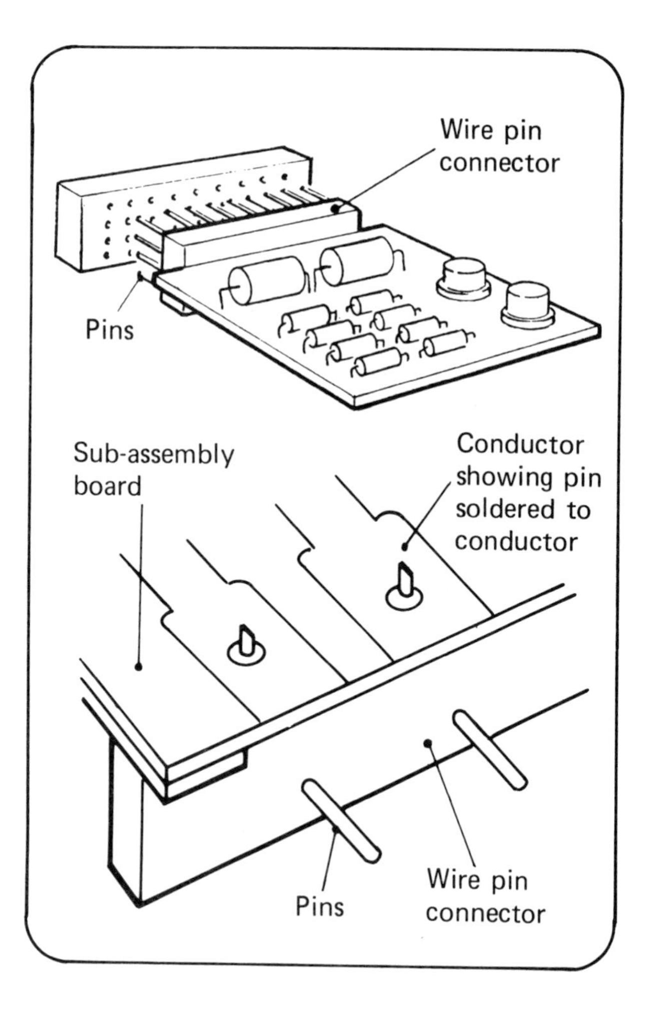

Wire pins

The body of the wire pin type connector is fixed firmly to the basic material of the board. The wire pins are soldered directly to the conductor, but the body allows no flexing or strain on the joint.

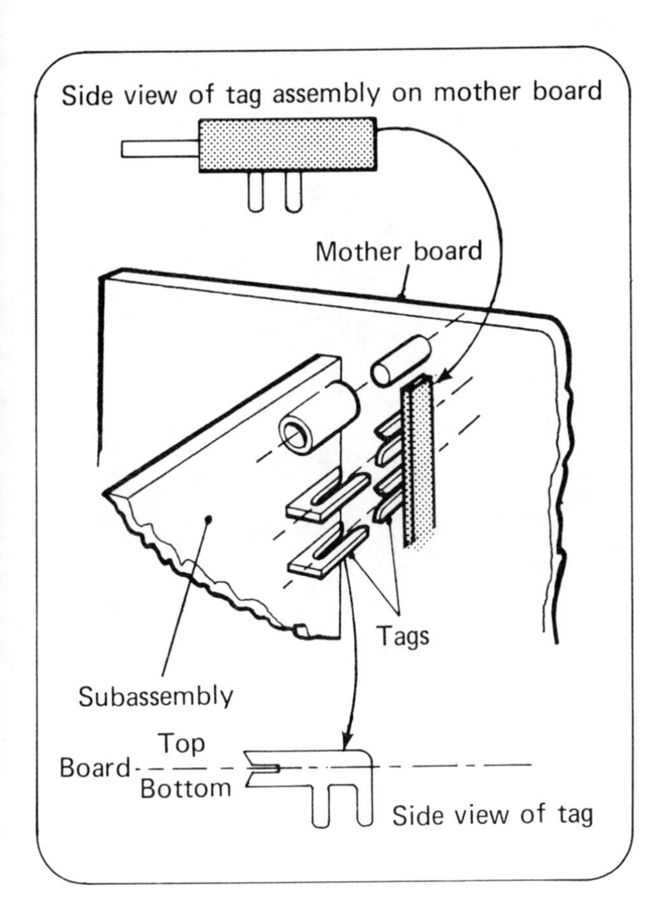

Tag connectors

These provide a separate removable tag for each circuit. Their extremely small size allows many circuit connections in a very small space, and this type of connector may be used in the mother/daughter board configuration.

Test items

1 Correctly describe the following types of connectors: edge, wire pin, and tag.

(See page 116.)

Module eleven
Encapsulation of PCBs

Introduction

This module examines a technique used to protect printed circuit boards from damage. Encapsulating a printed circuit board helps protect the board from vibration and other hazards such as liquids and dust.

Cleaning of printed circuit boards is often necessary once the board has been used in a piece of equipment. This module examines standard cleaning and handling procedures for printed circuit boards.

Key technical words

- **Encapsulation** Placing a protective coating over the components of a PCB.

Performance objectives

Upon completion of this module, you should be able to:

- State two reasons for encapsulating a printed circuit board.
- Name the four steps required to encapsulate a board.
- List three dangers of using a cleaning fluid.
- List two reasons why an assembled PCB should be packed in a protective cover.

Encapsulation of PCBs

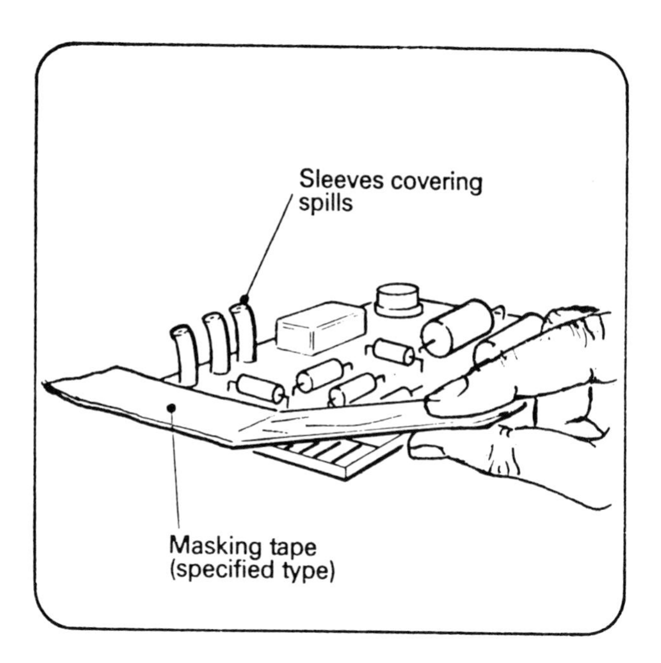

An encapsulated PCB is able to withstand vibration and other environmental hazards to a greater extent than an untreated board. When the covering is set, it holds all the components firmly in place and protects them from damage. The encapsulating compound is an insulator and all plug, socket, and spill connectors must be masked at the beginning of the process. Only the specified type of masking tape should be used.

Clean the board with an approved cleaning agent. Paint the board with the correct primer using a soft brush. Cover all surfaces except contacts to insure that the encapsulation compound makes a good bond. The primer must be the correct one for the type of board and for the compound being used.

After the compound has been mixed, it is painted onto the board. All surfaces must be covered to the recommended depth. The job is then left to set and harden.

> ***Note:*** The compound, after mixing, remains workable for only a certain time, depending upon the particular chemicals used. Before using already mixed encapsulate, check how long it has been setting.

After the board covering has set, place the board in the encapsulation tank and bake.

> ***Note:*** Check that you are working to the manufacturer's process specifications for the compound that you are using.

Remove protective coverings

When cured, remove the PCB from the tank and check that the surface is smooth, well-covered, and unbroken. Carefully remove the masking from the contacts and the sleeves from the spills.

Cleaning the board

Cleaning of PCBs

A PCB should be kept clean. This reduces risk of contamination damage. The first cleaning method involves the use of a brush and a specified cleaning fluid. Gently brush along the conductor and around all the components. Some fluids may affect a particular type of board or components; check that you have the correct cleaning fluid.

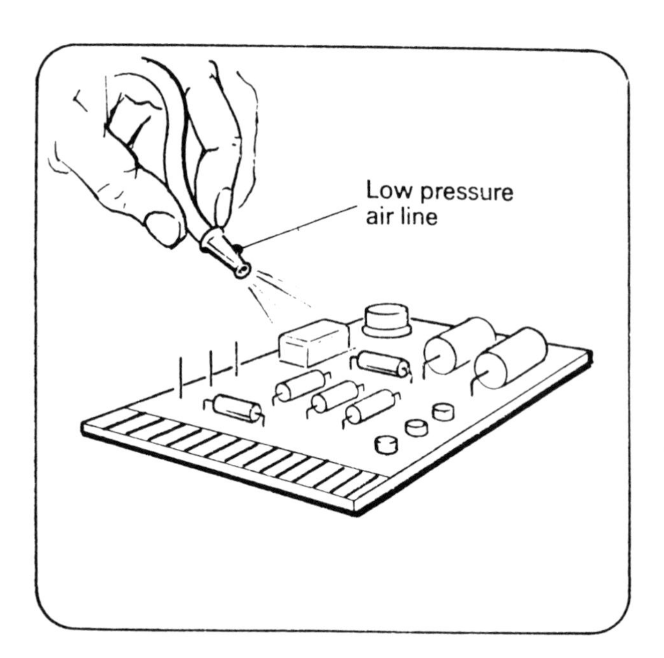

Foreign matter and dust may be removed with a low pressure air hose. Make sure the air pressure is not too high for the type of components you are using.

Never clean a PCB with a cloth or coarse brush—if you do, you may damage the conductors.

Safety

Compressed air is dangerous. Use carefully. *Never* direct it towards any part of the body.

Safety

Some cleaning fluids are very dangerous

(a) In a confined space.

(b) When in contact with heat.

(c) When inhaled, particularly through a cigarette.

DO NOT SMOKE.

Use the correct fluid.

Read and comply with all Safety Regulations.

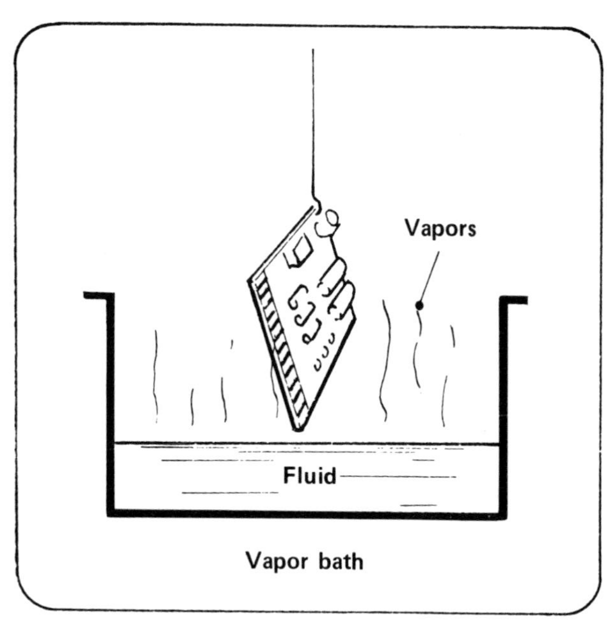

Vapor bath

It may be necessary to clean the board in a vapor bath. Check that it is the right type for your board. Read the safety regulations before attempting to use the bath. The board should be cleaned by the vapor and should not enter the fluid.

Prevention of damage to PCBs

Treat PCBs with care; they can easily be damaged by rough handling. Do not bend or twist them; the conductor may lift and the board may fracture.

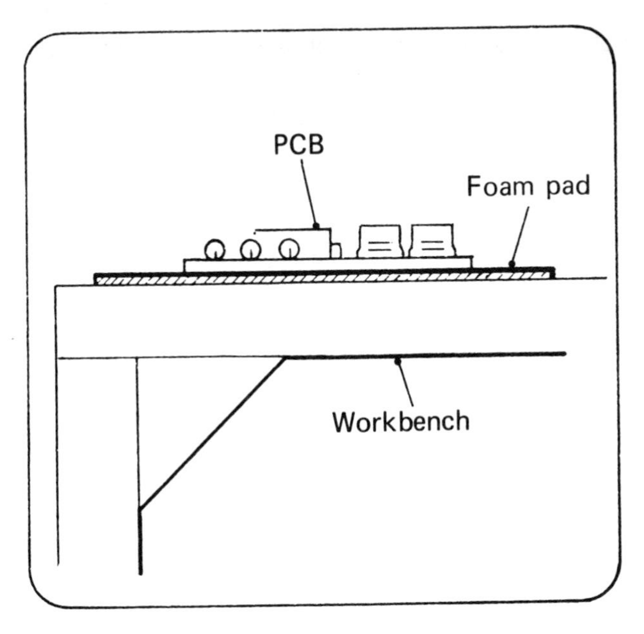

When working on a PCB, rest the conductor on a pad of foam rubber or similar material, or rest the component side on a material that will prevent excess force being exercised on any component.

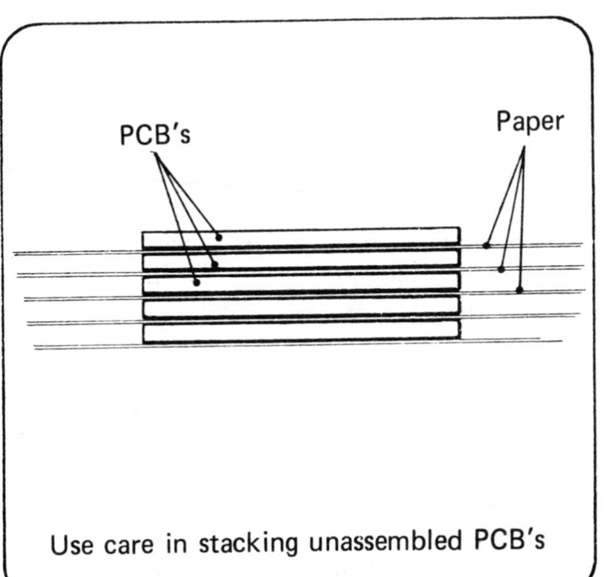

Use care in stacking unassembled PCB's

When stacking unassembled boards, place a piece of tissue paper of the specified type between each board, with the bottom board on a foam pad, or place each board in a polyethylene bag.

Quality

Contact with certain materials including some types of paper can cause damage to a PCB. (For example, some papers contain acid.) Use only acceptable and specified materials.

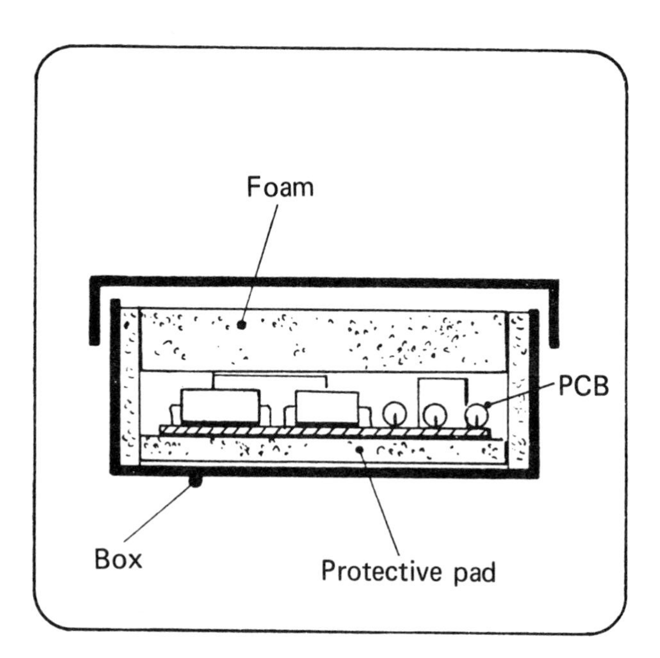

When moving or storing a complete board assembly, pack it in a protective cover with sufficient foam to prevent static charge damage or crushing.

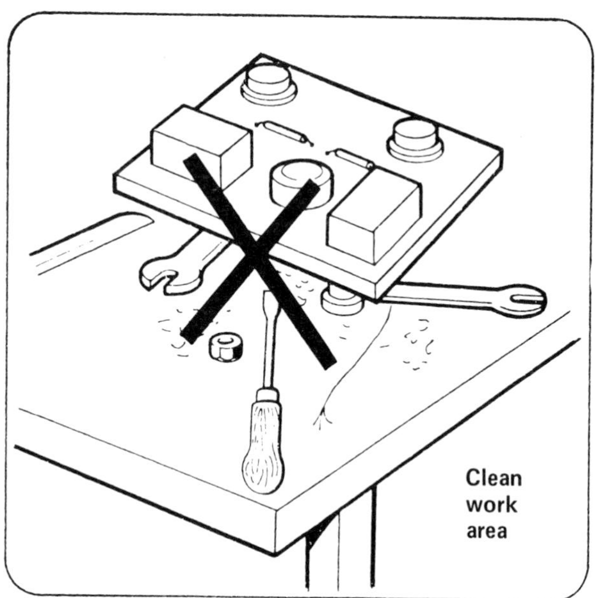

Do not allow a board or any delicate electronic equipment to rest on tools or other hard objects. Keep benches clean and uncluttered.

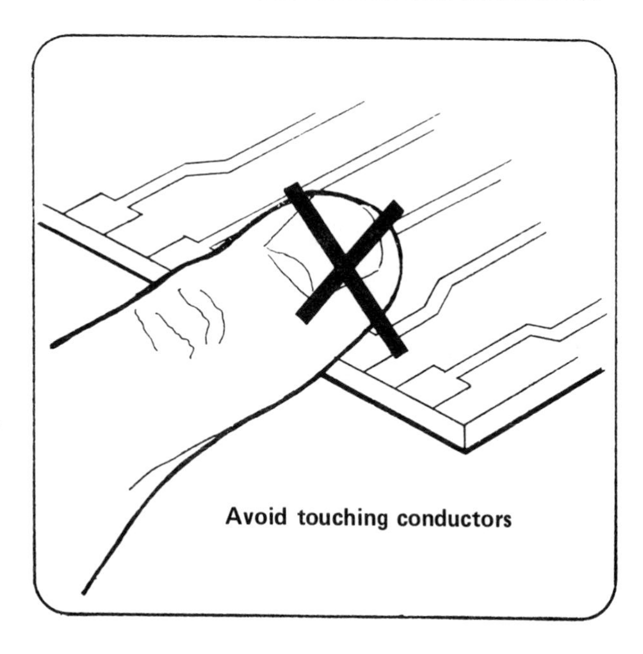

Contamination can cause a board to be rejected. Touch the conductor as little as possible, and be sure that your hands are clean.

Quality

Certain hand creams and preparations contain substances harmful to PCBs. Check before using any preparation, including a barrier cream.

Quality

Excessive ultraviolet light can damage boards. Keep PCBs away from direct sunlight and any source of UV radiation.

Benches should be kept clean—many fluids, including coffee can cause damage to PCBs.

Clumsy use of tools or careless handling of the board can cause damage such as broken conductors. Always treat the boards with care. Before soldering, check that the soldering iron is of the correct type and wattage. An excessively hot iron can cause blistering and lifting of the conductor.

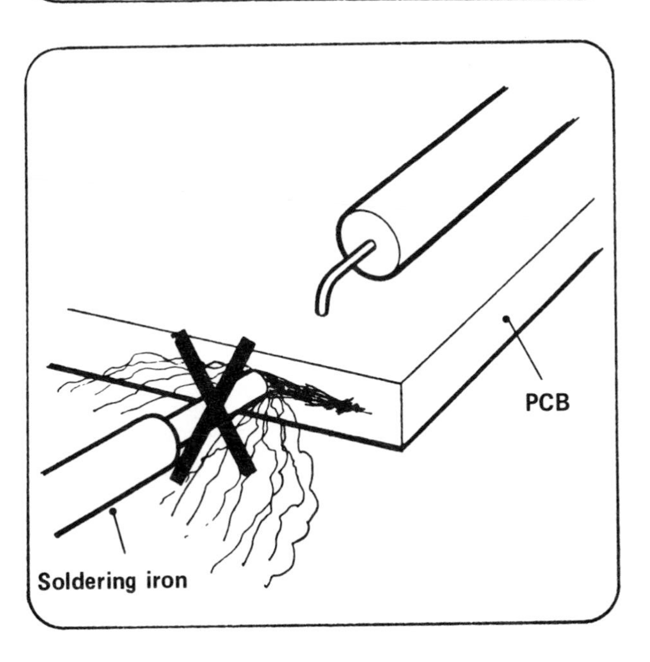

Soldering irons should be kept in their stands when not in use. When in use, care must be exercised to prevent damage to boards by irons held carelessly or left in the wrong place.

Test items

1 List two reasons why a PCB should be encapsulated.

(a) ______________________________

(b) ______________________________

(See page 120.)

2 There are four major steps required to encapsulate a printed circuit board. Fill in the missing two steps.

1. Clean the board with approved cleaner.

2. ______________________________.

3. ______________________________.

4. Bake the board.

(See page 120.)

3 Cleaning fluids are very dangerous. List three areas where extra precautions should be taken.

(a) ______________________________

(b) ______________________________

(c) ______________________________

(See page 122.)

4 List two reasons why an assembled PCB should be packed in a protective cover.

(a) ______________________________

(b) ______________________________

(See page 124.)

Module twelve
Repairs to PCBs

Introduction

This module examines basic repairs that can be made to a printed circuit board. Repairs can be made to conductors that become broken or that lift away from the boards. Conductor pads may also be repaired if they become damaged.

Boards that have been encapsulated require different techniques for repair. Basic repair procedures for encapsulated boards are reviewed in this module. The replacement of vertical and horizontal tag connectors may be necessary; this module examines the procedures for replacing tag connectors.

Performance objectives

Upon completion of this module, you should be able to:

- Describe how a broken conductor may be repaired using the bridge technique.
- Describe how a missing pad may be repaired.
- List the steps necessary to repair an encapsulated board.
- List the five steps for replacing a horizontal tag.

Repairs to PCBs

> **Quality**
>
> Before starting any repair to a PCB, check the specifications covering the particular type of boards, to determine if the type of repairs required are allowed.
>
> Some specifications forbid most repairs to PCBs. Under certain circumstances, for example, prototype works, repairs of the type shown may be permissible.

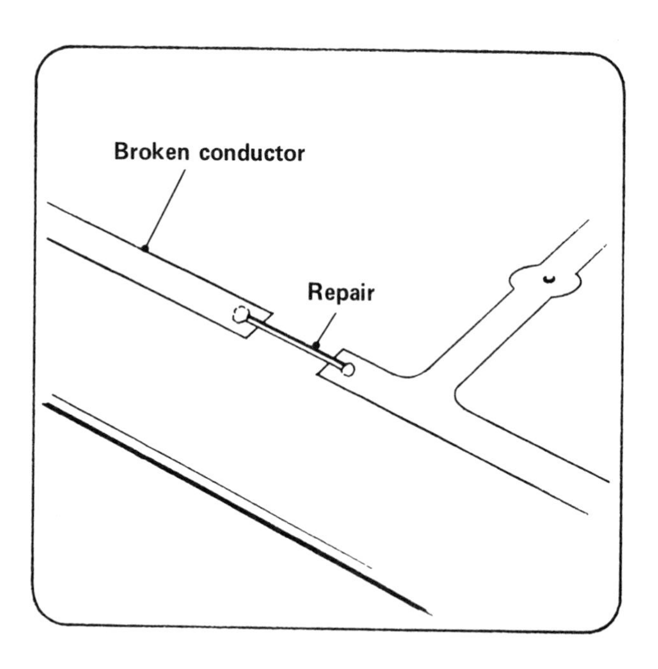

Broken conductor

Where a conductor has been broken but still remains firmly on the board, a strip of material similar to the conductor may be soldered across the break.

Clean the conductor and the bridge before soldering.

When cool, cover the joint with an acceptable varnish.

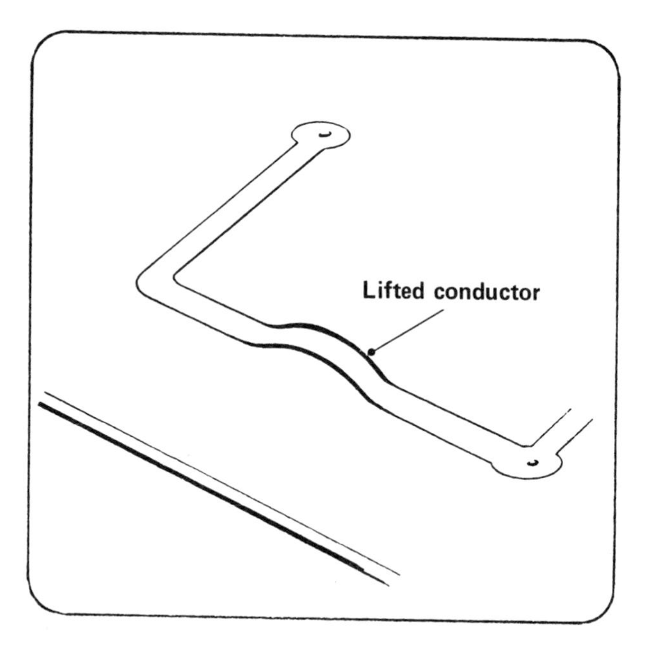

Lifted conductor

Where a conductor has lifted in one place but is otherwise adhering, it may be possible to effect a satisfactory repair.

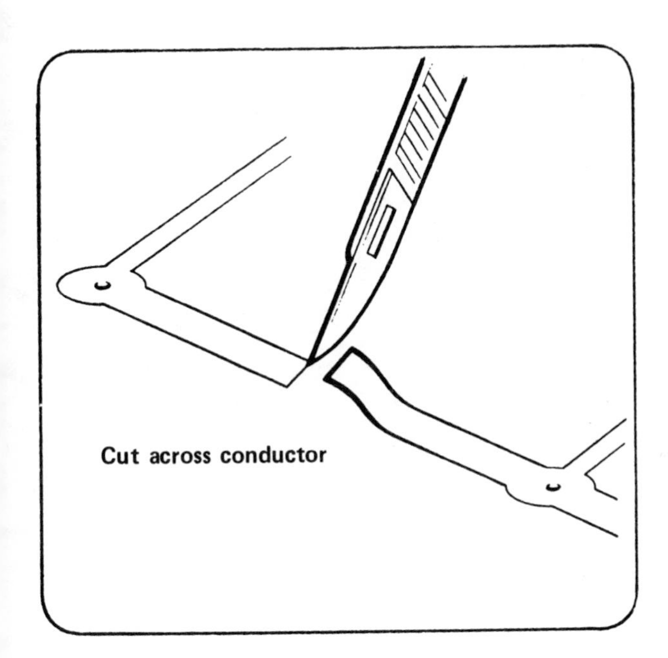

Clean the board and place it, component side down, on a piece of heavy rubber, plastic ring, or similar device so that the components will not be damaged while the board is being worked on. Carefully cut across the conductor on either side of the lifted portion at a point where adhesion is still good.

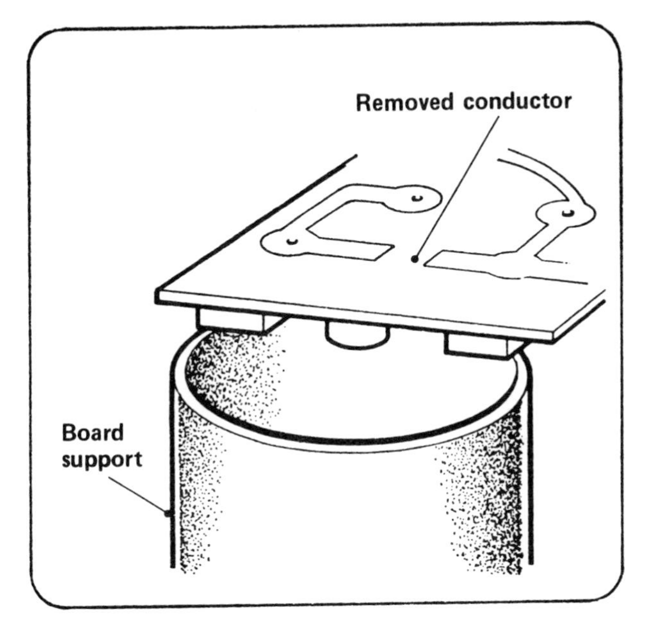

Remove the cut piece leaving a gap in the conductor.

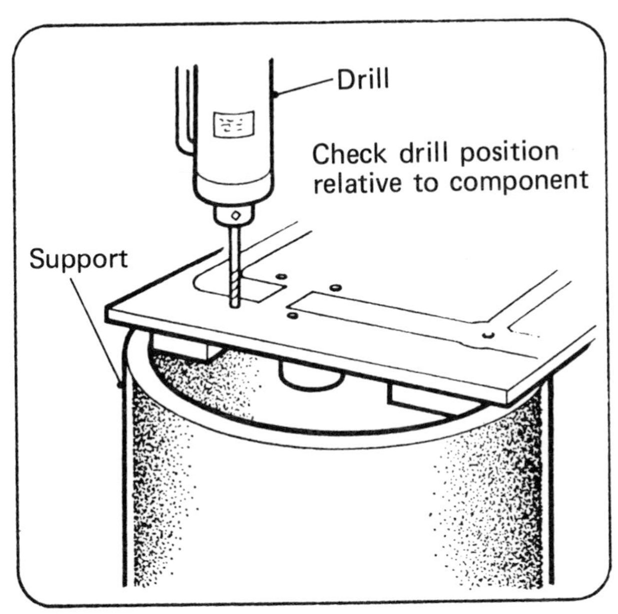

Drill four holes: two on each side of the break, one on each side of the conductor. Check their positions so that the drill will not damage components.

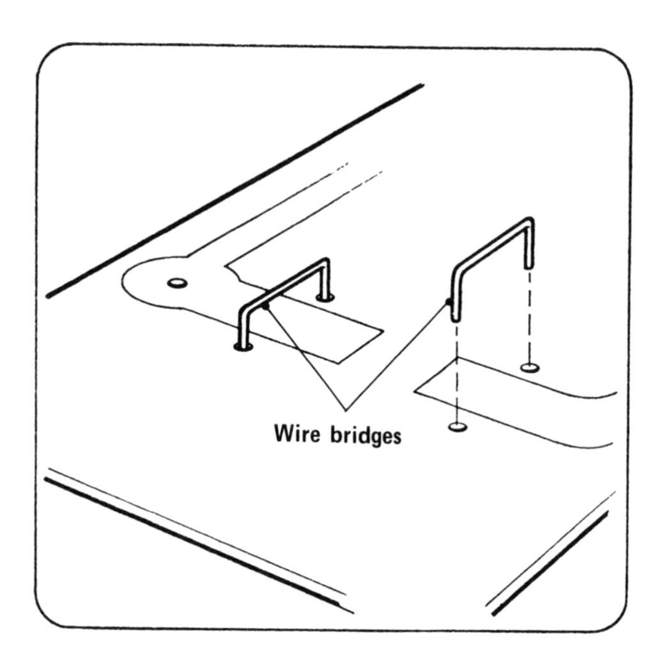

Accurately form wire bridges and insert them in the drilled holes.

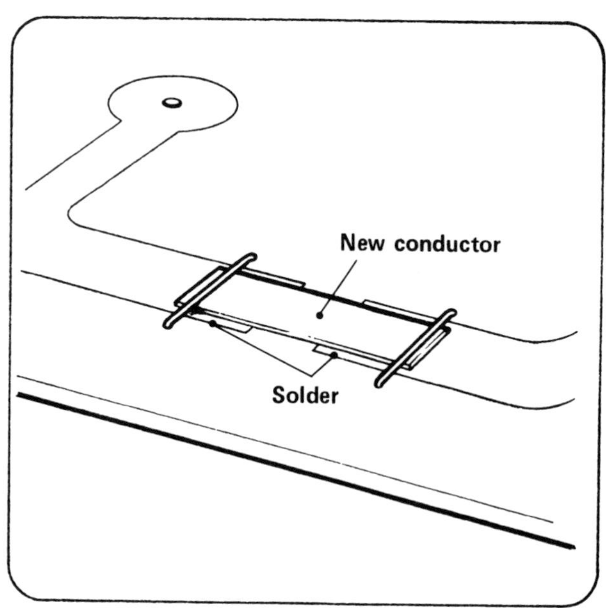

Insert the new portion of the conductor as shown. Bend the legs of the bridges on the component side of the board to hold the new portion of the conductor firm.

Solder bridges, new conductor, and old conductor. Check that

The conductor is firmly held.

There is adequate clearance between the repair and other conductors on the board.

The legs of the bridges on the component side are not in the way.

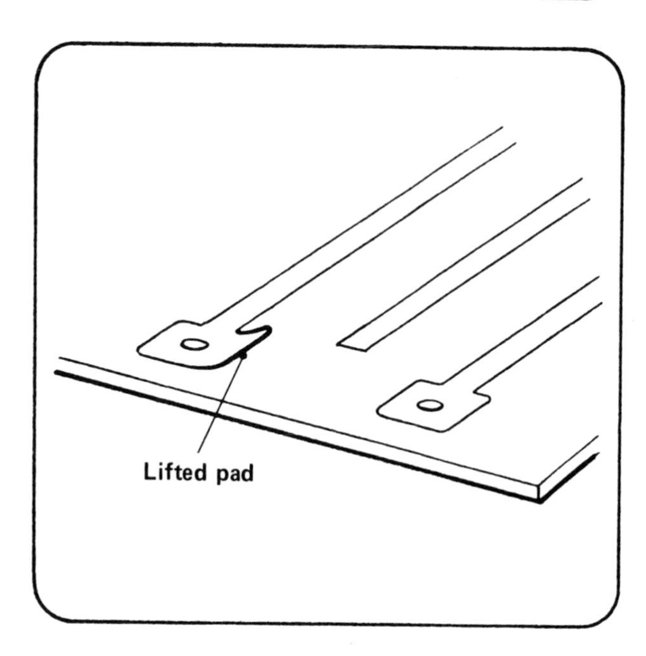

Damaged or lifted pads

Conductor pads sometimes lift and may break away. It is often possible to make a "good-as-new" repair.

Remove any component connected to the pad. Use a soft brush and a safe acceptable cleaning fluid to clean underneath the lifted pad, or the area of the missing pad. Fix the lifted pad with a suitable adhesive.

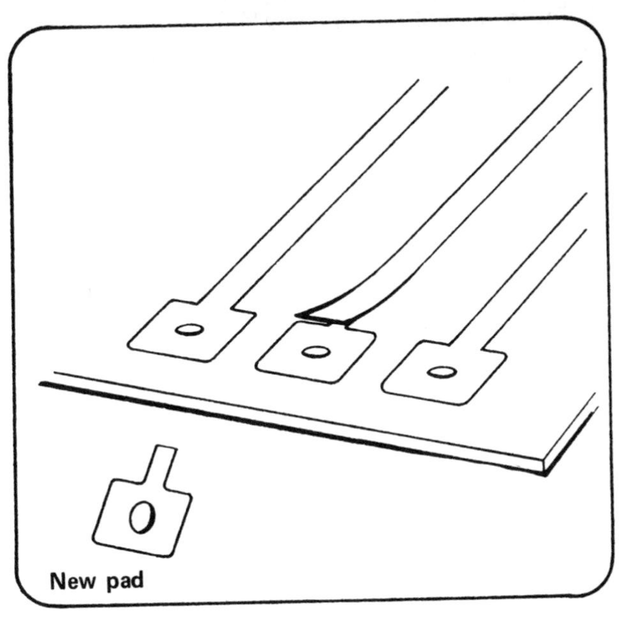

For a missing or broken pad, form from similar material to the original pad, a new pad having a longer tail than the missing piece. Using an acceptable adhesive, fix the new pad in place of the old, working the extra long tail under the remaining conductor.

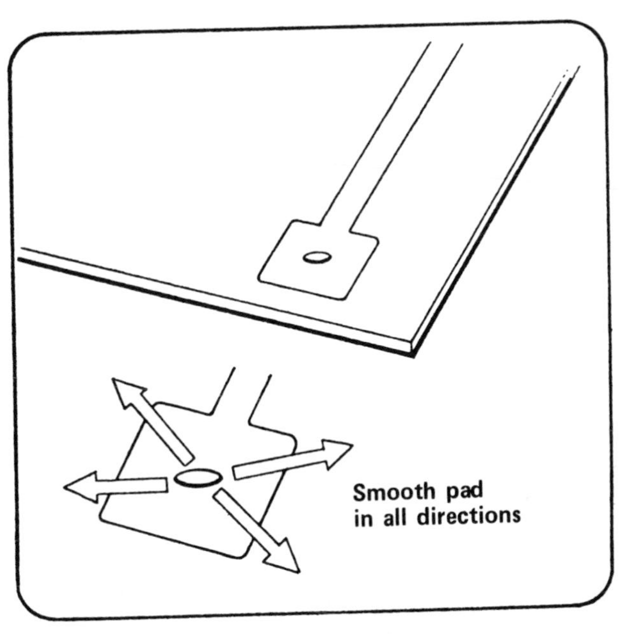

Work the pad gently and firmly to insure that no air is left under the track. When satisfied that the adhesive is binding over the whole surface, leave for 24 hours before soldering components and varnishing the joint.

Cut encapsulate around component

Components on an encapsulated board

Replacement of these components is applicable only where a pliable encapsulate has been used. Components on encapsulated boards may fail and will have to be replaced. The procedure is

1 Using a sharp blade make cuts in the encapsulate around the failed components. The cuts must be deep enough to break the seal, but not deep enough to damage the board.

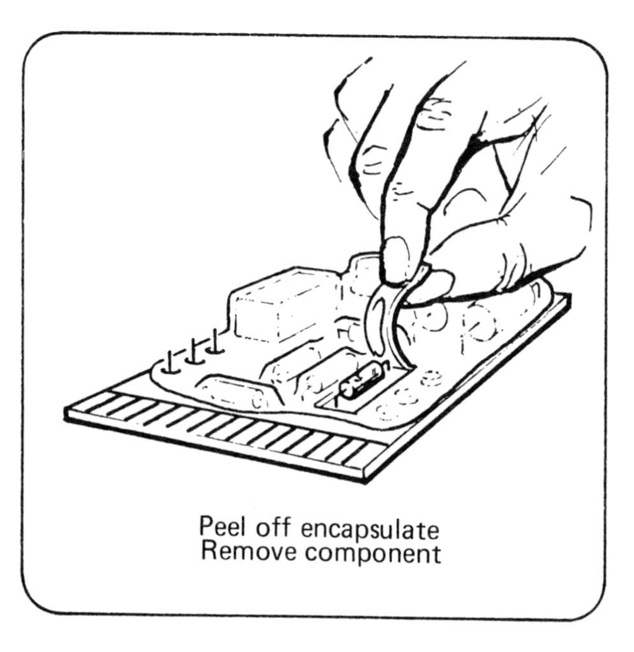
Peel off encapsulate
Remove component

2 Gently peel the encapsulate from the component, making sure that other components are not being stressed.

3 Remove the unserviceable component.

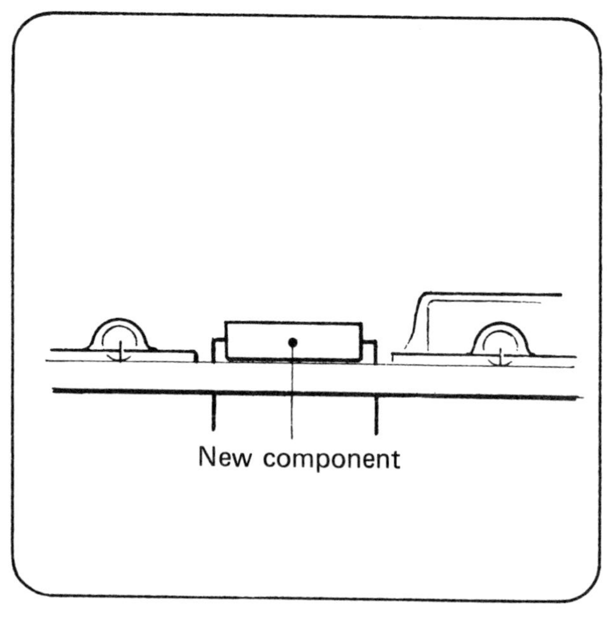
New component

4 Fit and solder the replacement component. Inspect and test before proceeding.

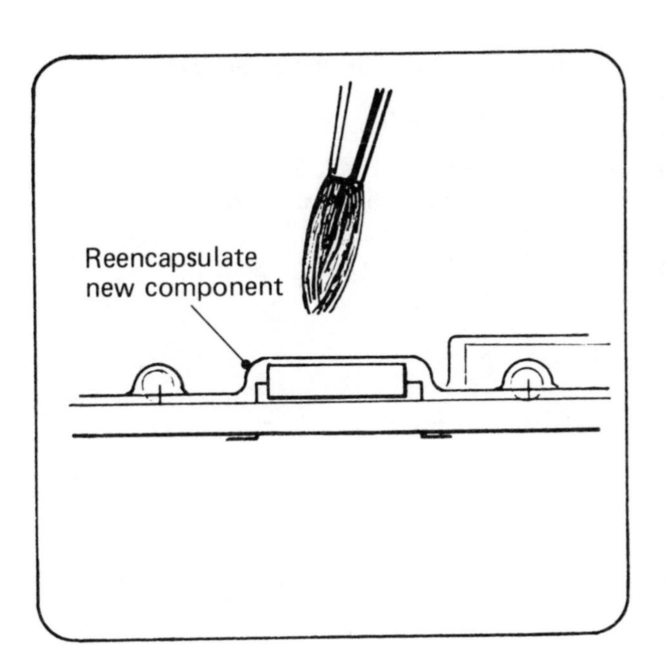

5 Reencapsulate the stripped area following the same procedure as for full encapsulation. Reinspect and test.

Replacement of tag connectors

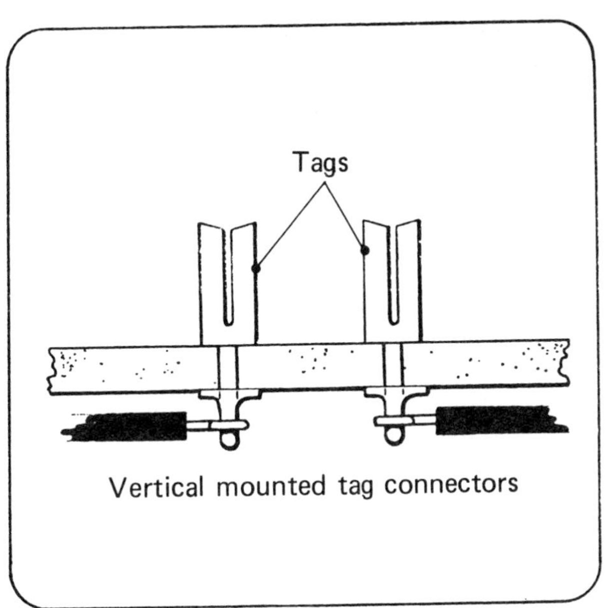

Vertical mounted tag connectors

Tag connectors are fixed to PCBs and a pair of tag connectors join at right angles to each other. One is mounted vertical to the board.

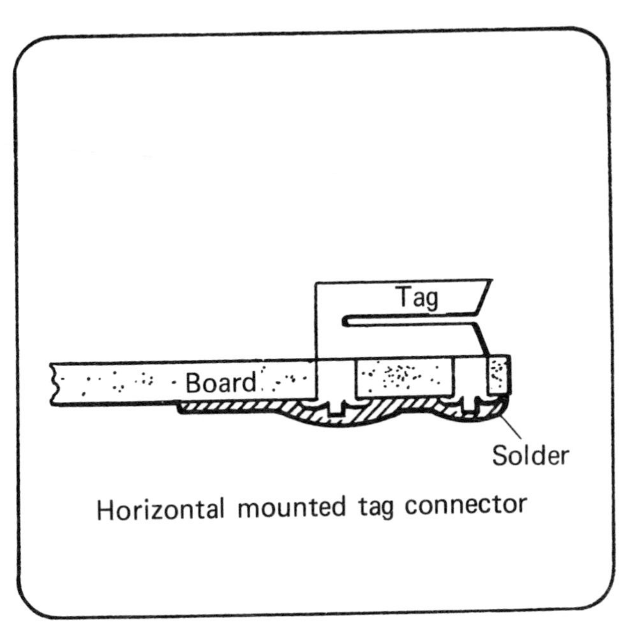

Horizontal mounted tag connector

The other board is mounted horizontal.

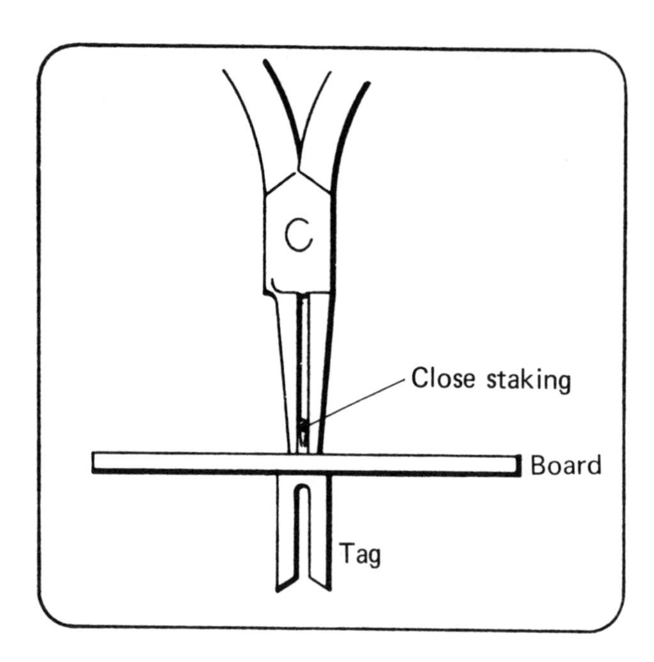

The procedure for removing a vertical tag is

1 Desolder the tag.

2 Close the staking with pliers, taking care not to damage the conductor or pad area.

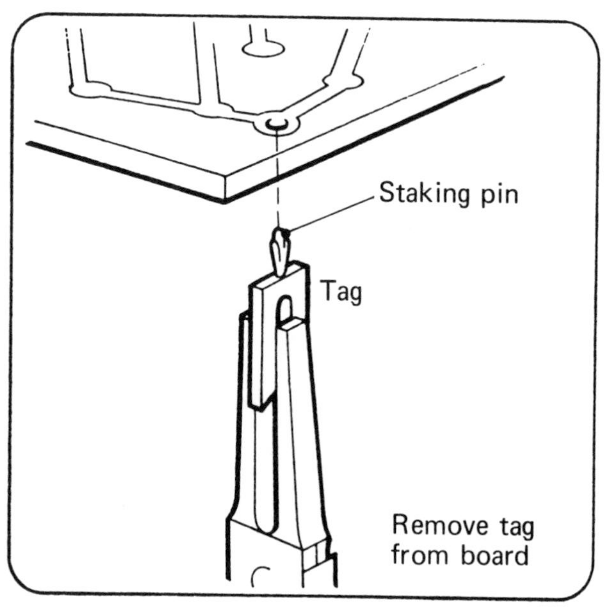

3 Using pliers, ease the tag out of the board with a straight pull.

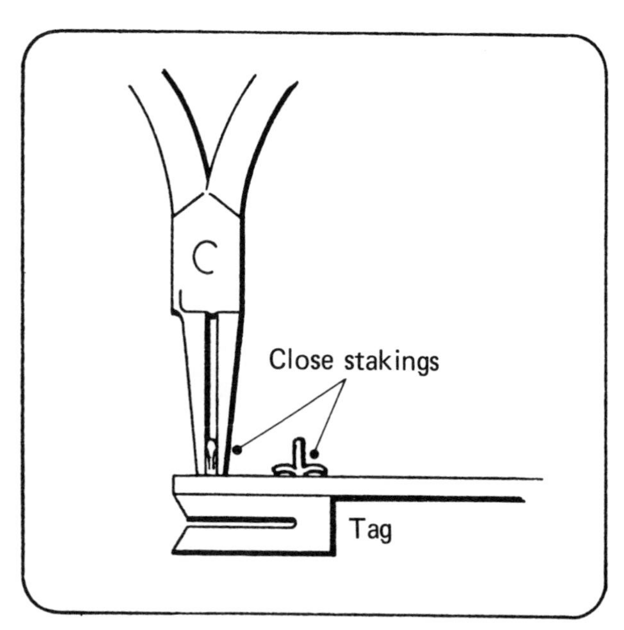

The procedure for removing a horizontal tag is

1 Desolder the tag.

2 Close the staking with pliers.

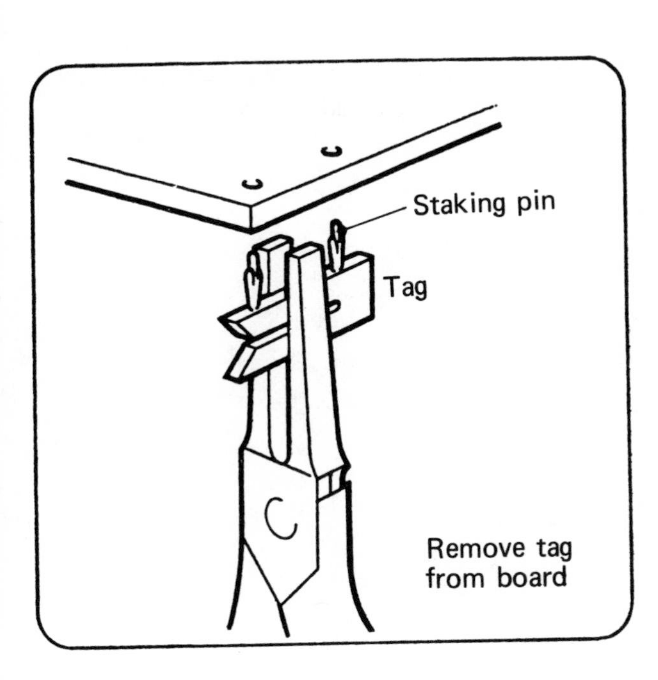

3 Ease the tag from the board.

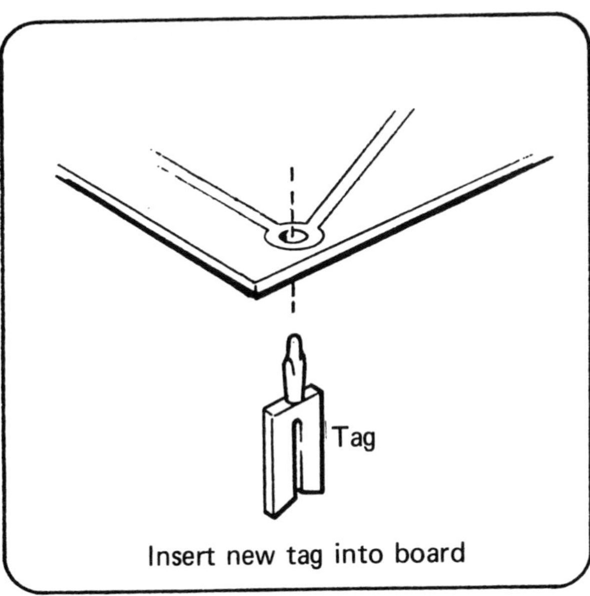

The procedure for replacing a vertical tag is

1 Insure that the hole in the board is clean and smooth.

2 Gently insert the shaft into the board.

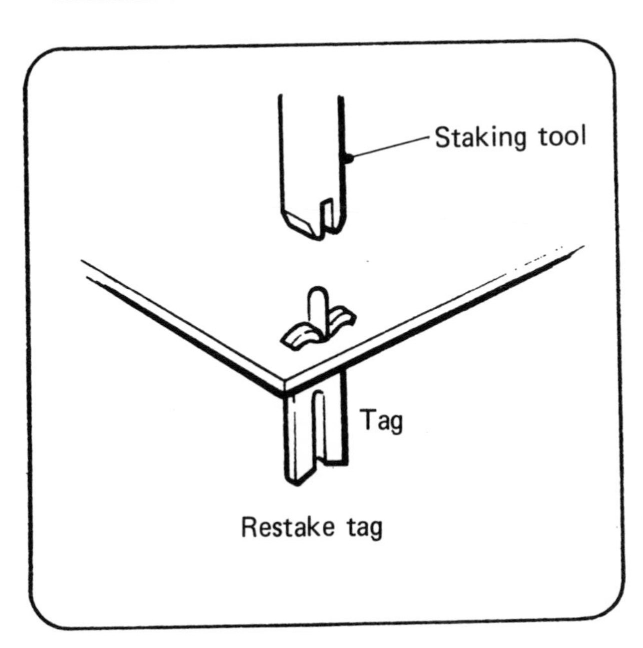

3 Restake as shown with a staking tool.

4 Carefully resolder leads and connections.

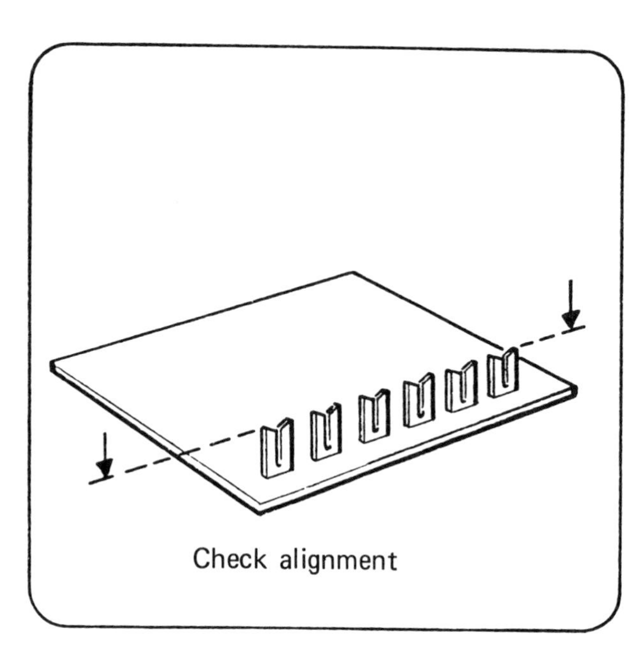
Check alignment

5 Check for alignment with other tags. A square and a straightedge can aid in alignment.

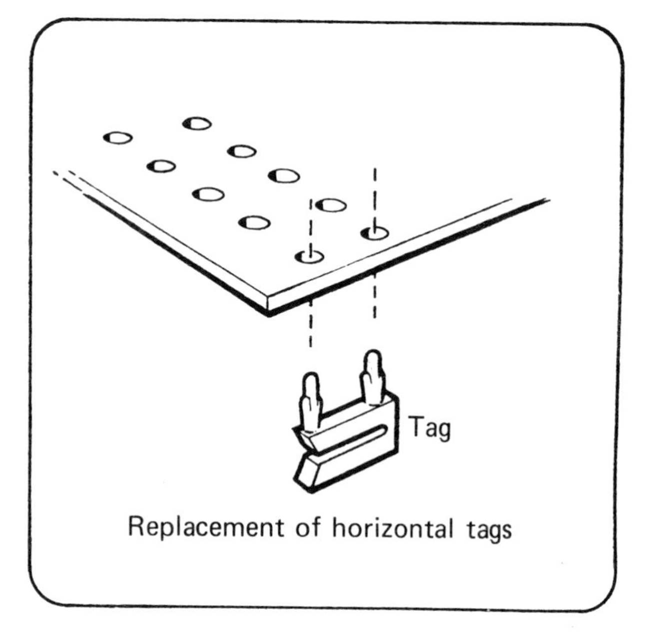

Replacement of horizontal tags

The procedure for replacing a horizontal tag is

1 Insure that the holes in the board are clean and smooth.

2 Gently insert the two pegs into the holes.

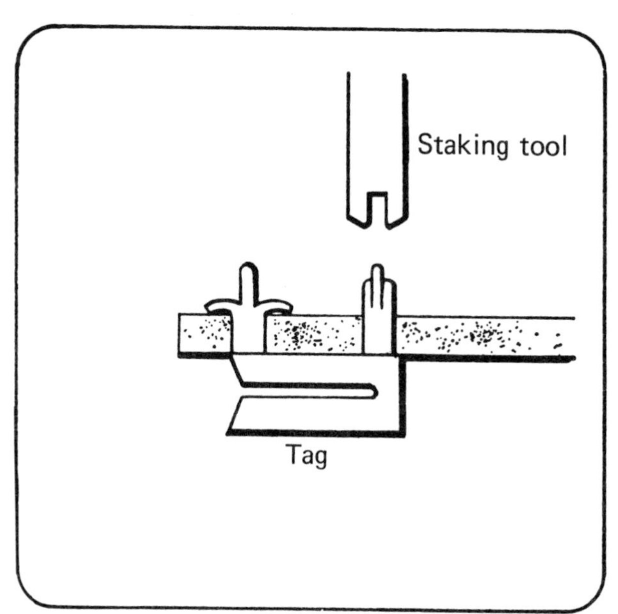

3 Restake as shown.

4 Resolder leads and connections.

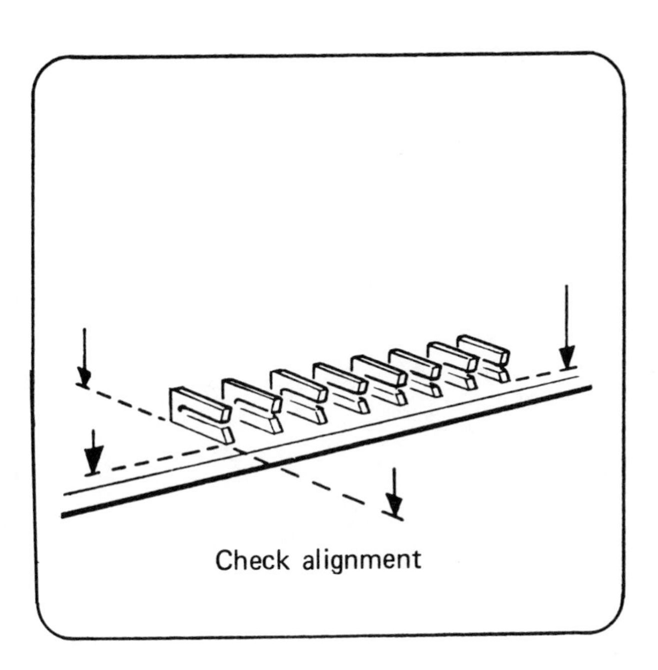

5 Check for alignment with other tags.

Test items

1 When repairing a printed circuit board, why should the component side of the board be placed on a piece of rubber or a plastic ring?

____

(See page 129.)

2 In using the wire bridge technique to repair a broken conductor, what should you check for before drilling the holes?

____

(See page 130.)

3 How would you remove the encapsulate from a faulty component on a printed circuit board?

____

(See page 132.)

4 Complete the following list of steps required to replace a horizontal tag.

1. Clean and smooth holes on board.
2. ______________________________.
3. Restake.
4. ______________________________.
5. Check alignment.

(See page 134.)

Module thirteen

Modifications to printed circuit boards

Introduction

This module examines the changing or modification of a printed circuit board when design changes are required. Modifications on printed circuit boards are very time-consuming and costly but may be required until new boards can be made. This module shows how spill tags or terminals may be used to modify existing boards.

Key technical words

- **Spill tag** A type of terminal that may be inserted into a PCB and then used to make a solder connection from one point on the board to another point on the board.

Performance objectives

Upon completion of this module, you should be able to:

- List the steps required to insert a spill tag.
- Determine the correct bit size for drilling the holes for the wire.
- Describe how far a spill tag should be inserted into the PCB.

Modifications to PCBs

Modifications involving conductor

Fixing of spill tags in PCBs

Spill tags are used for making connections between conductor and wire ended components, and as standoff terminals in boards. The type shown is a plain tapered tag with a serrated area. Tags are also called terminals.

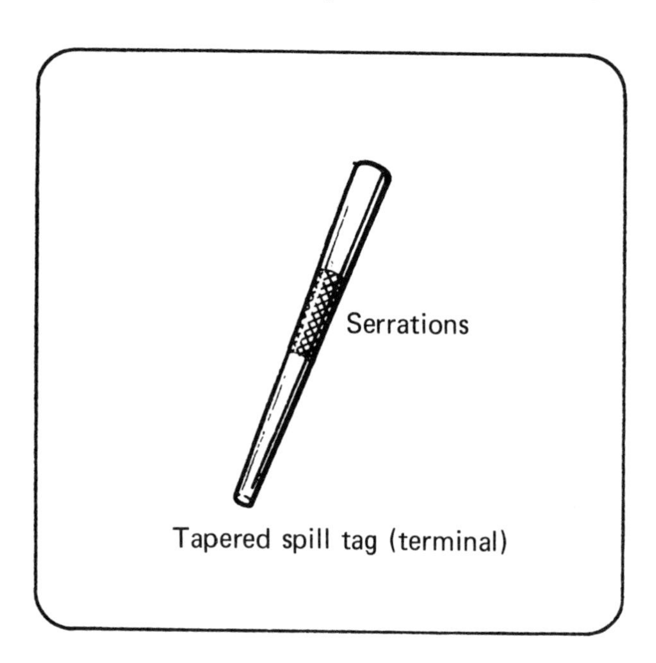

Tapered spill tag (terminal)

1 Drill a small hole for the spill tag between two conductors, or to the side of a single conductor.

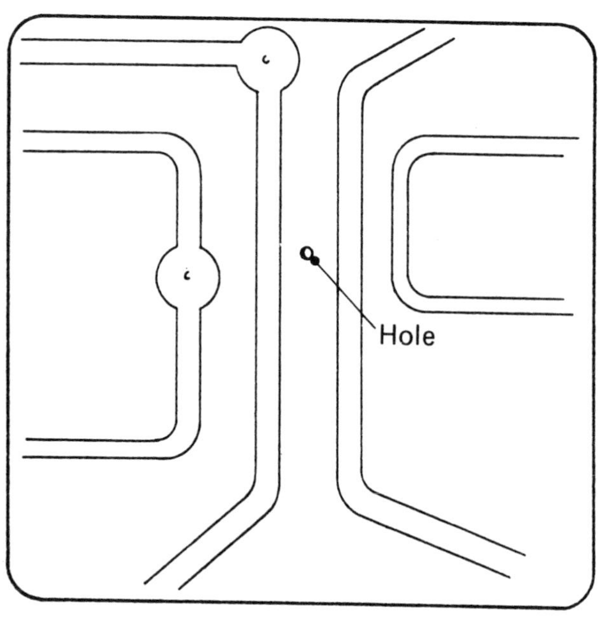

2 Select a very fine drill whose diameter is slightly larger than the diameter of the wire to be used as a link between the conductor and the spill. (For 28 awg wire, a 0.5 to 0.6 mm drill bit or a number 76 to 73 drill bit should be used.)

3 Drill a hole carefully through the center of the conductor adjacent to the first hole.

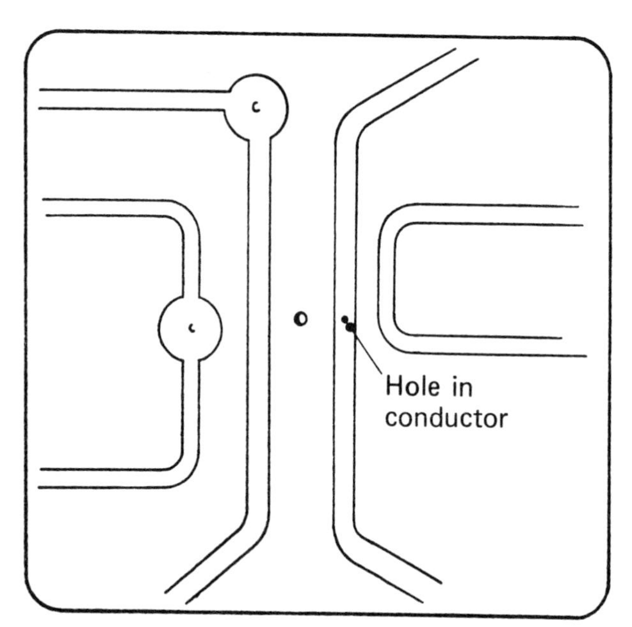

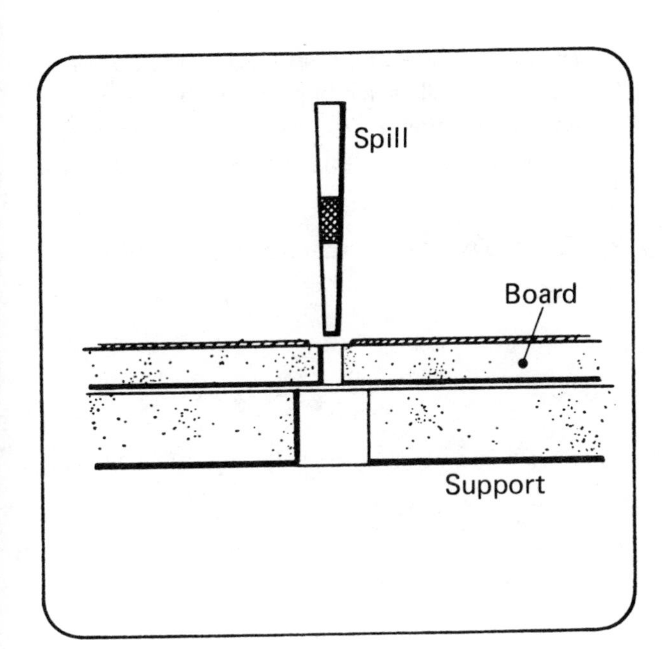

Support the work, leaving the area around the hole clear. Insert the spill into the larger hole from the conductor side.

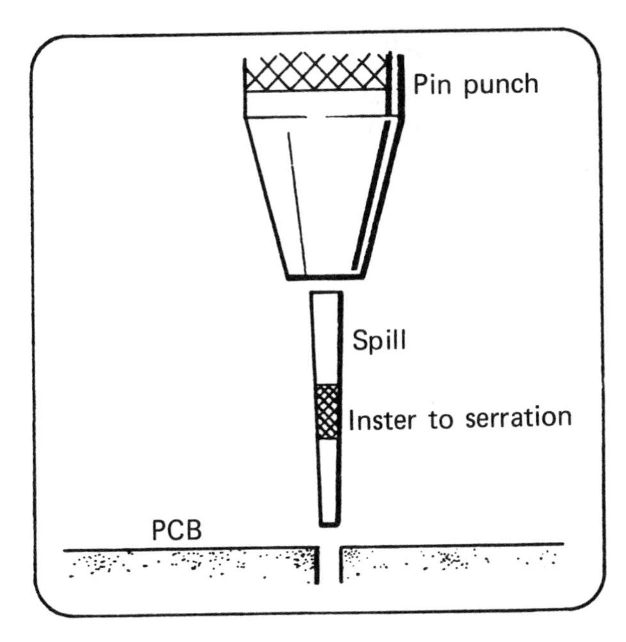

Tap the head of the spill with a pin punch. Keep the punch in line with the spill to avoid bending the spill or cracking the board. Insert the pin up to the serrations.

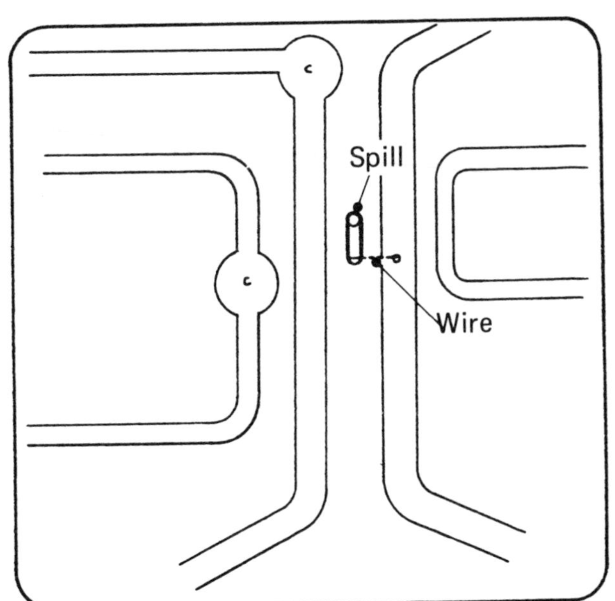

1 Form a mechanical joint around the pin with 28 awg tinned copper wire.

2 Lay the wire against the board.

3 Pass the wire through the hole in the conductor.

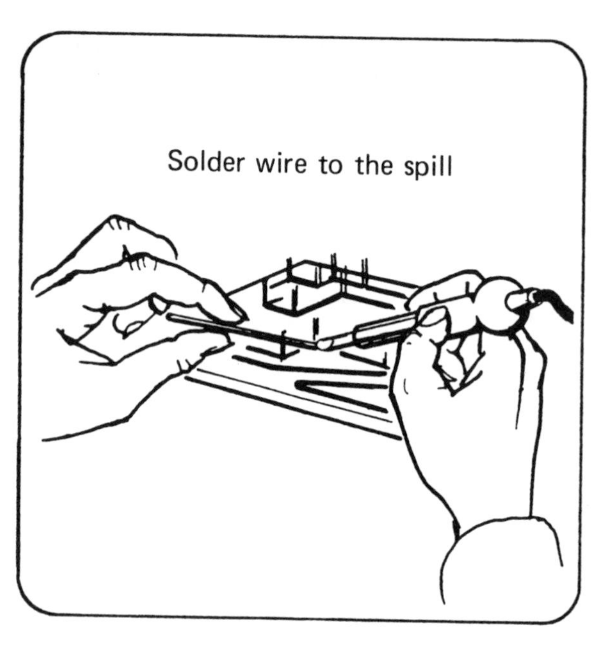

1 Solder the wire to the spill. Allow the solder to flow slowly to allow air held by the serrations to escape. If the solder covers too quickly, the trapped air will blow back and cause pin holes or prevent a good joint.

2 Trim the wire and solder it to the conductor.

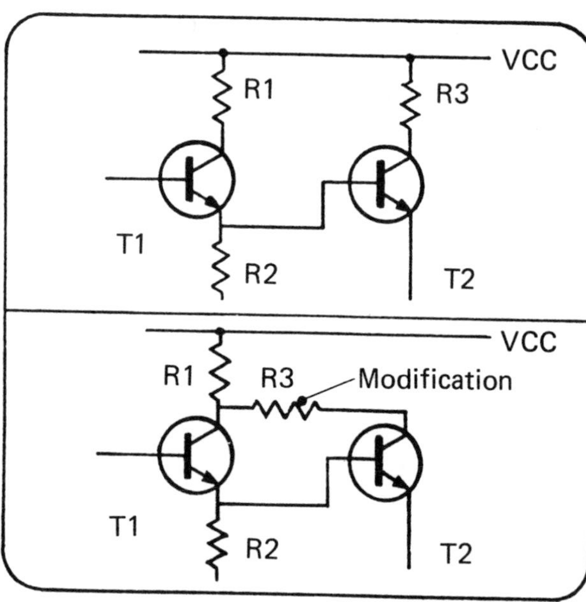

Circuit modifications

When a circuit is changed or modified, it may be necessary to alter the printed circuit before newly designed boards are available. The example shows the unmodified circuit on top, with the modified connection of resistor R3 in the bottom detail.

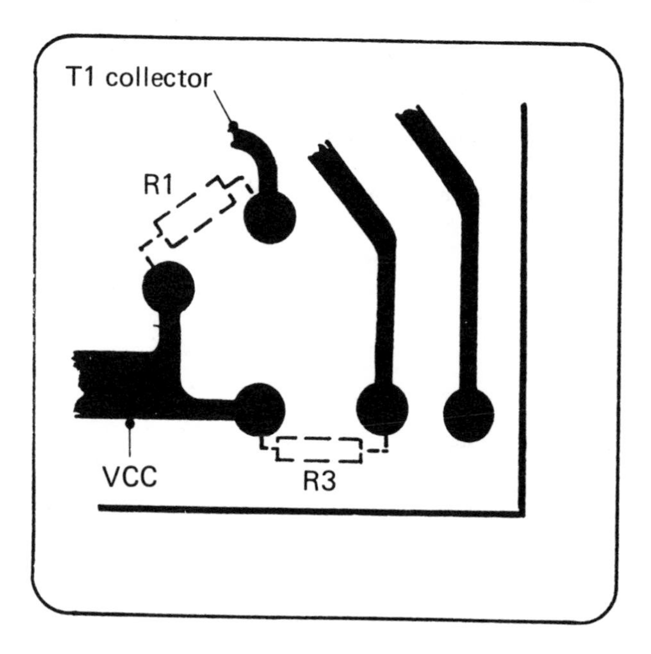

The existing printed circuit does not suit the new circuit configuration.

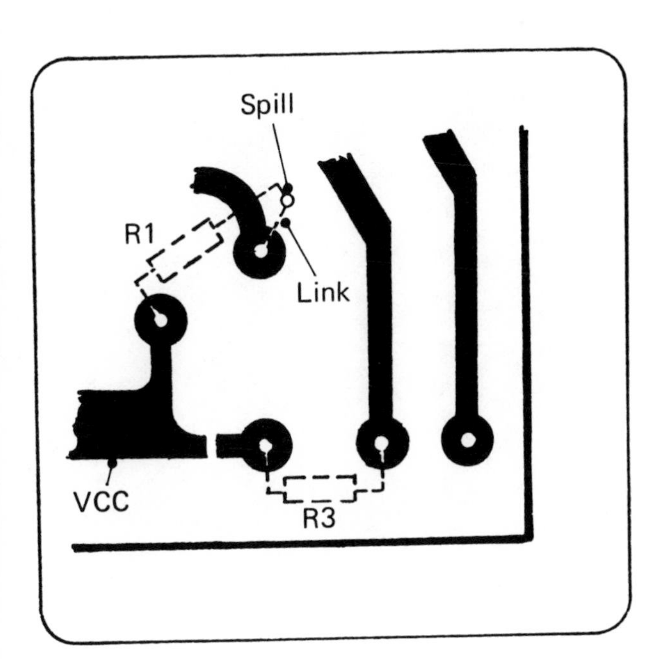

When the new layout of components has been derived from the amended circuit diagram, a new arrangement for the printed circuit can be seen.

The PCB is modified by cutting and removing a small section of the conductor and by inserting spills and soldering these to the conductor with wire links as necessary. Spills are inserted as shown but may be inserted in the conductor hole if sufficient area is available.

Connect tinned copper links between the spills and adjacent pad as shown.

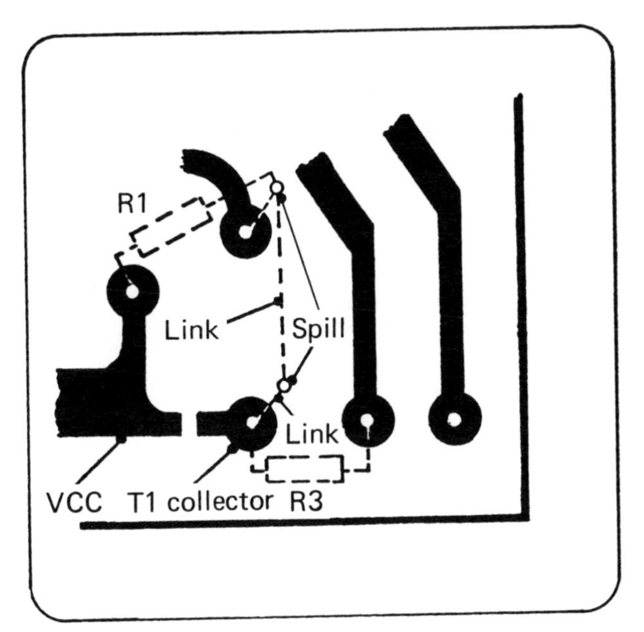

The components are resoldered and a wire connection is made on the component side of the board between the spills.

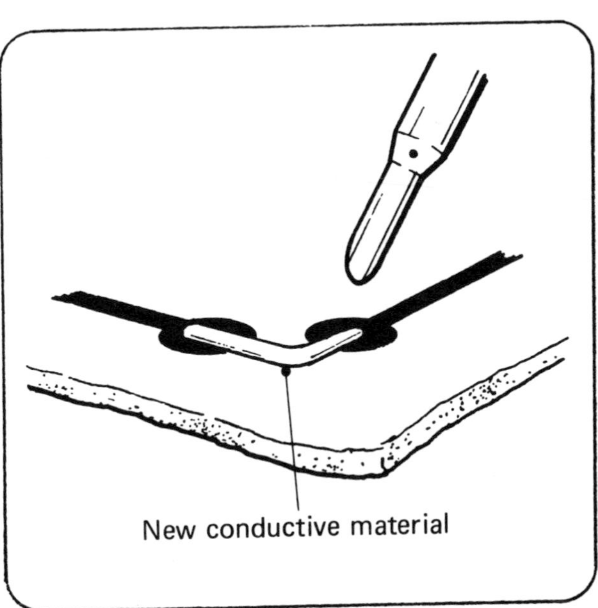

Where a modification embodies only an extra connection in the conductor, join the two points of the conductor by a piece of conductor material. Solder the material directly to the conductor and when the material is cool, varnish with an approved nonconductive varnish.

Test items

1 How far should a tapered spill tag be inserted into the board?

____

(See page 141.)

2 If a wire link is made from 28 awg wire, what size drill bit should be used to drill the holes?

____

(See page 140.)

3 Complete the following list of steps that should be used when a spill tag is used to modify a PCB.

1. Drill a small hole for the spill tag.

2. ____.

3. Insert the spill tag into the board.

4. ____.

5. Solder the wire to the spill.

(See page 142.)

Module fourteen
Re-flow soldering

Introduction

This module explains one of the final stages in the mass production of printed circuit boards, re-flow soldering. The fusion or re-flow process of soldering components has several advantages that are identified in this module. The soldering process is a machine-soldering process instead of hand-soldering. The re-flow technique is used for miniature and microelectronic printed circuit boards.

Key technical words

- **Re-flow or fusion process** A manufacturing technique used for printed circuit boards where the conductor and the components are tinned before they are soldered together.

Performance objectives

Upon completion of this module, you should be able to:

- State the advantages of the re-flow technique.
- State what should be checked on the re-flow soldering machine before the machine is used.

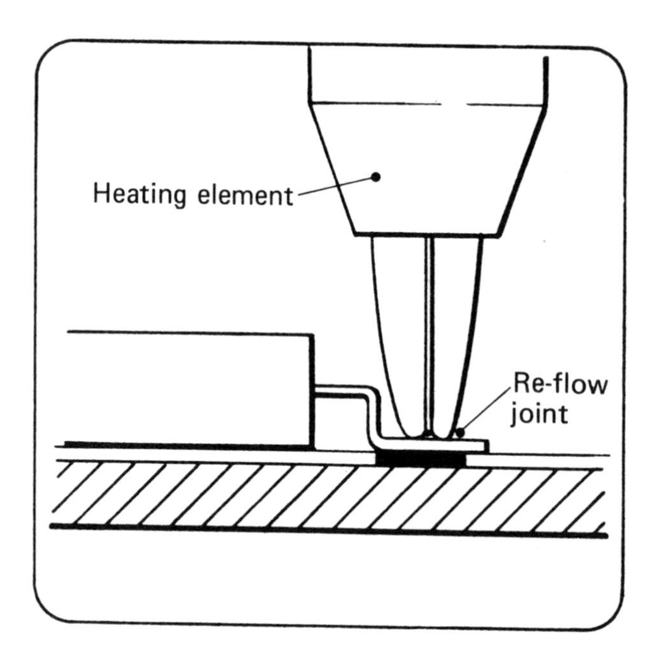

Re-flow soldering

Machine re-flow soldering is particularly applicable to soldering conductors, especially miniature and microminiature modules. The two surfaces are tinned. They are then placed together by a machine that accurately controls the time and temperature of the soldering process.

The re-flow soldering process is also called "fusion" and has several advantages.

1 Improved solderability.

2 Extended shelf life—the copper surfaces are plated.

3 A quality control step—it shows if copper was properly cleaned.

Current is passed into the heating elements of the machine for a predetermined time, the solder tinning flows, and a good joint is obtained. Pressure is maintained until the element and the joint have cooled. Then the machine will lift the heating element automatically.

When setting a re-flow soldering machine, insure that the pressure is set as specified: too little pressure will allow air in the joint, too much will force the solder from between the two surfaces.

Test items

1 List three advantages of using the re-flow soldering technique.

(a) ______________________________

(b) ______________________________

(c) ______________________________

(See page 146.)

2 In order to obtain a good solder joint, the re-flow soldering machine's _______ should be set to the correct specifications before beginning the soldering process.

(See page 146.)

Index